3rd International Conference on Civil Engineering and Architecture (ICCEA 2020)

Held online due to COVID-19

IOP Conference Series: Materials Science and Engineering
Volume 910

Compiegne, France
28 - 30 June 2020

ISBN: 978-1-7138-3165-5
ISSN: 1757-8981

Printed from e-media with permission by:

Curran Associates, Inc.
57 Morehouse Lane
Red Hook, NY 12571

Some format issues inherent in the e-media version may also appear in this print version.

Institute of Physics
Dirac House, Temple Back
Bristol BS1 6BE UK

Phone: 44 1 17 929 7481
Fax: 44 1 17 920 0979

techtracking@iop.org

Additional copies of this publication are available from:

Curran Associates, Inc.
57 Morehouse Lane
Red Hook, NY 12571 USA
Phone: 845-758-0400
Fax: 845-758-2633
Email: curran@proceedings.com
Web: www.proceedings.com

TABLE OF CONTENTS

DESIGN AND ANALYSIS OF BUILDING MATERIALS

EARTHQUAKE AND STRUCTURAL ENGINEERING

ARCHITECTURAL DESIGN AND PROJECT MANAGEMENT

ENVIRONMENTAL PROTECTION AND SPATIAL PLANNING

IOP Conf. Series: Materials Science and Engineering **910** (2020) 011001 doi:10.1088/1757-899X/910/1/011001

Preface

Dear Authors and (Online) Participants of ICCEA 2020,

On behalf of the conference committees, I am pleased to declare that we had really successful three days of 2020 3rd International Conference on Civil Engineering and Architecture (ICCEA 2020), which was supposed to be held in Compiègne, France, during June 28-30, 2020, but was changed to be held online due to the COVID-19 pandemic. The conference was sponsored by Seoul National University and the University of Technology of Compiègne. Again, the online conference was of great success, totally beyond anybody's expectation, having a diverse group of expertise and backgrounds. Not to mention that participants are geographically diverse. I am glad to confirm that such diverse expertise and backgrounds were connected, forming a coherent conference environment and creating synergistic learning and sharing.

ICCEA is the premier forum for the presentation of new advances and research results in the fields of theoretical, experimental, and practical civil engineering and architecture. The conference indeed brought together leading researchers, engineers and architects in the domain of interest from around the world. ICCEA 2020 was composed of 4 oral parallel sessions (15 min. for each talk, including 2-3 min. for Q&A), 4 keynote speeches (45 min. for each keynote speech, including 5 min. for Q&A) delivered, respectively, by myself; Prof. Adnan Ibrahimbegovic, the University of Technology of Compiègne, France; Prof. Youngjin Lee, Boston Architectural College & Sasaki Associates, Inc., USA; and Prof. Bernhard Egger, Seoul National University, South Korea; and finally, I would like to thank all of our conference committee members and participants for always being supportive to the conferences and joining online during their busy schedule to share valuable knowledge with us (and despite the time zone difference). I also would like to thank Prof. Dario Pedrabissi, the University of Portsmouth, UK, for his willingness to serve as a Program Chair of ICCEA 2020. As expected, the conference with all the keynote speeches and presentations was proven to be intellectually stimulating to all.

I just regret that we could enjoy a beautiful and historically significant city, Compiègne, and the beautiful campus of the University of Technology of Compiègne, France, but will get together in 2021 in another glorious place. We really enjoyed the conference, the hospitality (from the host, Prof. Adnan Ibrahimbegovic), and the constructive online discussion, and will continue to reinforce this strong foundation!

July 15, 2020

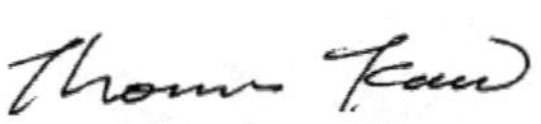

Dr. Thomas Kang
Professor, Seoul National University, South Korea
Conference Co-Chair, ICCEA 2020
tkang@snu.ac.kr

IOP Conf. Series: Materials Science and Engineering **910** (2020) 011002 doi:10.1088/1757-899X/910/1/011002

Conference Committee

Conference Chair

Thomas Kang, Seoul National University, South Korea

Adnan Ibrahimbegovic, University of Technology of Compiégne, France

Program Chair

Bernhard Egger, Seoul National University, South Korea

Youngjin Lee, Boston Architectural College & Sasaki Associates, Inc., USA

Dario Pedrabissi, University of Portsmouth, UK

Technical Committee

Raudhah Binti Ahmadi, Universiti Malaysia Sarawak, Malaysia

Ryan Ahn, Texas A&M University, USA

Yong Han Ahn, Hanyang University, South Korea

Mohammed Alajmi, Kuwait University, Kuwait

Majid Ali, Capital University of Science and Technology Islamabad, Pakistan

Allan Alzona, National University - Manila, Philippines

Lyazat Aruova, L.N.Gumilyov Eurasian National University, Kazakhstan

Satawat Doungpan, Sukhothai Thammathirat Open University, Thailand

Usama Ebead, Qatar University, Qatar

Emina Hadzalic, University of Sarajevo, Bosnia

Hyeon-Jong Hwang, Hunan University, China

Seung Yup Jang, Korea National University of Transportation, South Korea

Daranee Jareemit, Thammasat University, Thailand

Muneerah Jeludin, Universiti Teknologi Brunei, Brunei

Mohammad Arif Kamal, Aligarh Muslim University, India

Tamara Kelly Kelly, Abu Dhabi University, UAE

Manish Kewalramani, Abu Dhabi University, UAE

Jong R. Kim, Nazarbayev University, Kazakhstan

Sunkook Kim, Kyung Hee University, South Korea

Woosuk Kim, Kumoh National Institute of Technology, South Korea

Yun-Yong Kim, Chungnam National University, South Korea

Chompoonut Kongphunphin, Thammasat University, Thailand

Sashi Kunnath, UC Davis, USA

Seung-Jun Kwon, Hannam University, South Korea

Hung-Jen Lee, National Yunlin University of Science and Technology, Taiwan

Kang Seok Lee, Hanyang University, South Korea

Leonardo Massone, University of Chile, Chile

Mazlina Zaira Mohammad, Universiti Teknologi MARA, Malaysia

Ilwoo Nam, Nanjing Tech University, China

Grit Ngowtanasuwan, Mahasarakham University, Thailand

IOP Conf. Series: Materials Science and Engineering **910** (2020) 011002 doi:10.1088/1757-899X/910/1/011002

Zhmagul Nuguzhinov, Karaganda State Technical University, Kazakhstan
Edward Jonathan Soto Oblea, Universidad Peruana de Ciencias Aplicadas, Peru
Yu-Chen Ou, National Taiwan University, Taiwan
Alberto G. Perera, Architect & Planner, Spain
Chaiwat Riratanaphong, Thammasat University, Thailand
Lidyana Binti Roslan, Universiti Malaysia Sarawak, Malaysia
Samaila Saleh, Hassan Usman Katsina Polytechnic, Nigeria
Junwon Seo, South Dakota State University, USA
Shideh Shadravan, University of Oklahoma, USA
Kyung-Jae Shin, Kyungpook National University, South Korea
Manat Srivanit, Thammasat University, Thailand
Varodom Suksawasdi, Thammasat university, Thailand
Ian F. Taberner, Boston Architectural College, USA
Manny Anthony Taguba, National University - Manila, Philippines
Xiang Guo Wu, Harbin Institute of Technology, China
Haryati Yaacob, Universiti Teknologi Malaysia, Malaysia
Junyan Yang, Southeast University, China
Yanto Yanto, Jenderal Soedirman University, Indonesia
Wardah Fatimah Mohammad Yusoff, National University of Malaysia, Malaysia
Mohd Zairul, Universiti Putra Malaysia, Malaysia
Ning Zhu, Tsinghua University, China

Peer review statement

All papers published in this volume of *IOP Conference Series: Materials Science and Engineering* have been peer reviewed through processes administered by the proceedings Editors. Reviews were conducted by expert referees to the professional and scientific standards expected of a proceedings journal published by IOP Publishing.

Analysis of seismic bidirectionality on response of reinforced concrete structures with irregularities of l-shaped plan and soft story

V H Sobrado[1], R Yaranga[1] and J D Orihuela[1]

1Department of Civil Engineering, Universidad Peruana de Ciencias Aplicadas (UPC), Surco, Lima, Peru.

Email: u201517214@upc.edu.pe

Abstract. The seismic design of buildings is usually performed using one-way analysis for each of main axes independently. However, seismic events have fairly random behaviour and impose bidirectional solicitations on structures. In this work, the study of the response in structures subjects to earthquake loads with irregularity of l-shaped plan and soft story is carried out. For this, the linear time-story analysis (LTHA) of these has been carried out imposing seismic solicitations in two orthogonal directions. Thus, the structural response with incidence angle variations of 10° is obtained and compared with the response derived from the unidirectional analysis. Variations of up to 50% and 72% are obtained for model structures with l-shaped plan and soft story respectively.

1. Introduction

There are many countries that are exposed to the risk of a large seismic event, especially those located in the Ring of Fire. History shows that these events have produced economic and human losses, for example, in Armenia 1999 (Colombia), Pisco 2007 (Peru), Maule 2010 (Chile), Salta 2015 (Argentina) and Muisne 2016 (Ecuador) [1]. The earthquake in Puebla 2017 (Mexico) caused the collapse of a building, not because of design problems or soil-structure interaction, but because of seismic directionality, since the response obtained were greater than those calculated with unidirectional seismic analysis [2]. Thus, it's necessary that the seismic design of buildings is carried out considering the nature of seismic events, in this way the behaviour of models will be closer that of actual structures.

A large part of seismoresistant design codes indicate that the study of structures must be carried out by means of unidirectional seismic analysis on each main axis independently; However, this consideration isn't correct, since seismic movements are two-dimensional phenomena in the horizontal plane and have a random behaviour [3]. An important question of structural dynamics is whether the answers calculated with the traditional analysis are like the most accurate answers obtained considering the seismic bidirectionality. An important question of structural dynamics is whether the response calculated with the traditional analysis is like the most accurate the response obtained considering the seismic bidirectionality. In this regard, Kostinakis et al. [4] studied the influence of the angle of incidence in symmetrical buildings relative to their structural axes. They noted that the maximum response doesn't depend on the angle in symmetrical structures with equal stiffness in the two orthogonal directions; instead, the angle is influential in buildings that have different stiffness in their structural axes. On the other hand, Magliulo et al. [5], Cantagallo et al [6], Armaloo and Emami [7] conducted studies of the seismic behavior of irregular structures in plane. Results show that the angle of incidence significantly influences the response of the structures. In turn, Heredia and Machicao [8] developed a study of the structural demand of torsionally rigid and flexible models. The increase in response is greater in torsionally flexible models and increases with the degree of irregularity. These investigations show that the response of structures is influenced by the seismic bidirectionality.

This study considers two types of three-level reinforced concrete (RC) structures, one with l-shaped plane irregularity and other with soft story irregularity. First, the Response Spectrum Analysis (RSA) is performed, according to the traditional process indicated by the seismic design codes [9-11]. Second, LTHA is performed considering seismic bidirectionality with incremental incidence angles of 10°. Finally, the variation in interest response obtained by both types of analysis is analysed.

2. Methodology

2.1. Analysis methodology

The development of the present investigation considers two types of analysis, one to analyse the structures considering the seismic unidirectionality and the other to carry out the study of the seismic bidirectionality. On the one hand, RSA is performed, considered within many seismoresistant design codes. This traditional method performs the analysis unidirectionally on the orthogonal x and Y axes independently using elastic pseudo-acceleration spectra and damping factor C-0.05 [12]. LTHA is performed in accordance with Equation 1 using the modal combination method and transient time functions, as it is suitable for seismic solicitations [13].

$$\boldsymbol{k}x(t)+\boldsymbol{c}\dot{x}(t)+\boldsymbol{m}\ddot{x}(t)=r(t) \tag{1}$$

Where:

$\boldsymbol{m}$, $\boldsymbol{c}$ and $\boldsymbol{k}$ represent the mass, viscosity, and stiffness matrix of the structure respectively.

$\ddot{x}(t)$, $\dot{x}(t)$ and $x(t)$ represent the acceleration, velocity, and displacement of the structure respectively.

$r(t)$ are external forces that occur according to the passage of time.

In this one bidirectionality is considered, so solicitations are imposed in two directions perpendicular at the same time. These solicitations are studied for incidence angles with variations of 10°. The minimum seismic records to be used are three; these must be corrected by signal filtering and baseline, further, scaled according to the seismic parameters of the zone [14]. In addition, the structural models to be used are RC, framed structural system, and levels between three to four floors.

2.2. Consideration of seismic bidirectionality on structure response

Bidirectionality influences design seismic responses and can be a source of structure collapse [2]. The seismic responses that are calculated are at a global level and then the following equation is presented which considers a factor of amplification by bidirectionality to the responses calculated by unidirectional analysis.

$$R_f=KR_0 \tag{2}$$

Where, R_f is the final response that considers the bidirectionality in one axis, R_0 is the response that is obtained from the unidirectional seismic analysis in each axis and the K factor depends on the type of irregularity of the building and the response to be obtained; these factors are presented in results.

3. Results

3.1. Structural Models

To validate this study, we will consider two structures whose configurations have l-shape plane (model 1) with a mezzanine height on the first level of 4 m and the following 3 m, and the other with soft story (model 2) with a height of the first level of 4.5 m and the following 3 m. These structures correspond to common buildings in Lima. Soil with shear wave propagation speeds between 500 m/s and 1500 m/s is considered. The structural elements of both structures have sections of 45x70 cm2 the columns and 30x60 cm2 the beams. The study models are shown in figure 1.

IOP Conf. Series: Materials Science and Engineering **910** (2020) 012001 doi:10.1088/1757-899X/910/1/012001

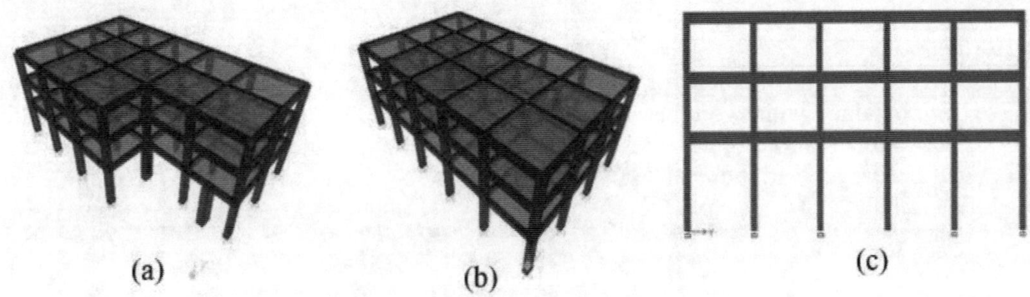

Figure 1. Model 1 with l-shaped plane irregularity (a), Model 2 with soft story irregularity (b) and Model 2 elevation (c).

3.2. Seismic solicitations
The following pseudo-acceleration spectra were used in RSA for the model and for each structural axis, this is shown in figure 2.

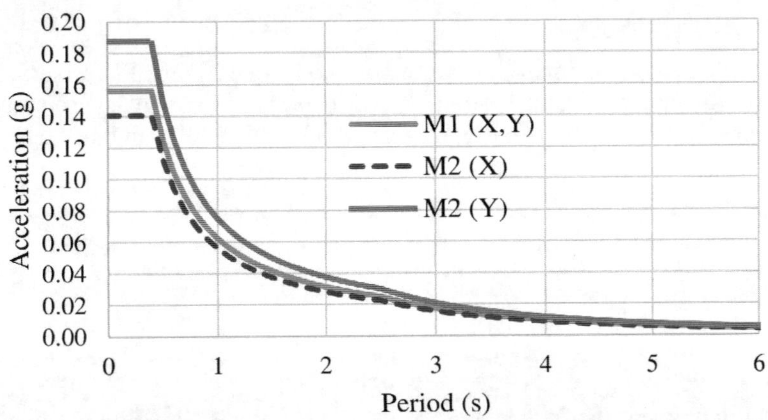

Figure 2. Pseudo-acceleration Response Spectrum

In LTHA, three seismic records of the Parque de la Reserva station, Cercado de Lima, Lima (latitude: -12.70, longitude: -77.04) are used. The following figure 3, figure 4 and figure 5 show the mentioned accelerograms:

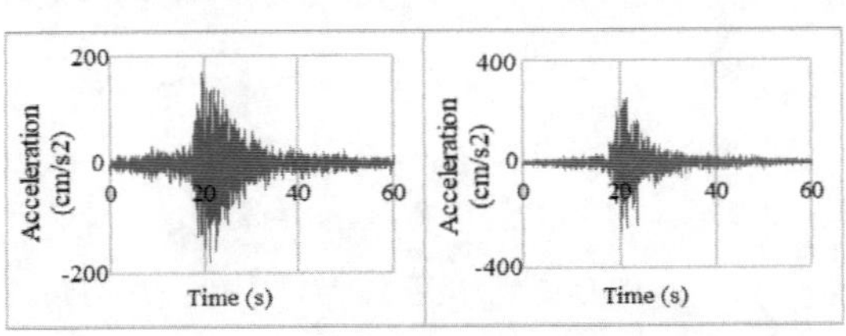

Figure 3. Accelerogram 1996 (EW, left and NS, right)

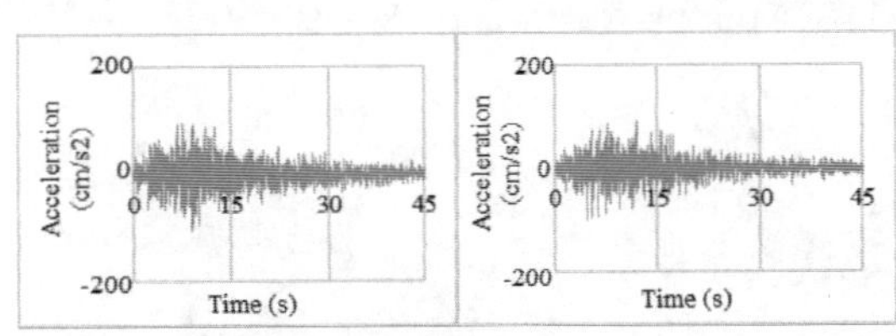

Figure 4. Accelerogram 1970 (EW, left and NS, right)

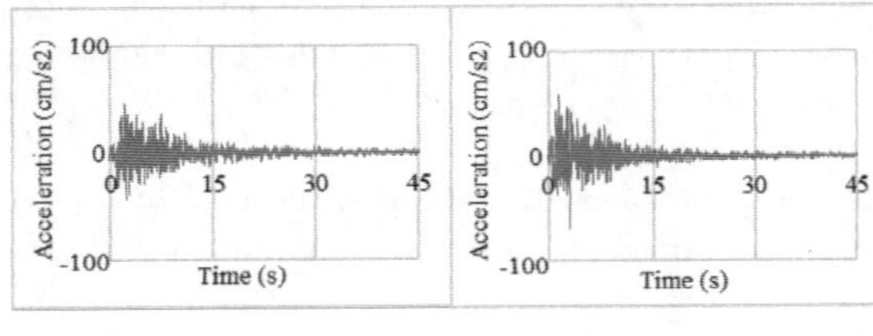

Figure 5. Accelerogram 1974 (EW, left and NS, right)

3.3. Basal shear

Figure 6 and figure 7 show variations of the basal shear as a function of the seismic incidence angle. These have been obtained with two types of analysis (RSA and LTHA). In model 1, it's observed that in the X axis there are variations of up to 31.61% (0°) and in the Y axis there are variations up to 49.67% (40°). In model 2 there are variations of 72.39% (130°) in the X axis and 18.77% (10°) in the Y axis.

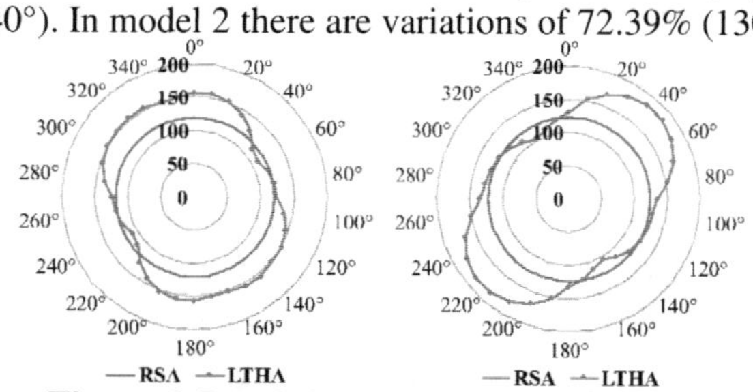

Figure 6. Basal shear (t) of Model 1 (X-axis, left and Y-axis, right).

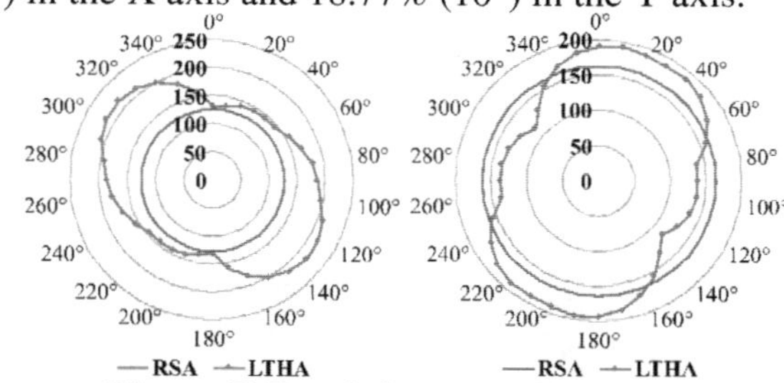

Figure 7. Basal shear (t) of Model 2 (X-axis, left and Y-axis, right)

3.4. Bending moment

Like the basal shear, the bending moment is analysed. In model one (figure 8), variations of up to 45.91% (40 °) and 32.15% (130 °) were calculated on the X and Y axes respectively. Also, in model 2 (figure 9) there were variations of 20.99% (10 °) on the X axis and 69.24% (130 °) on the Y axis.

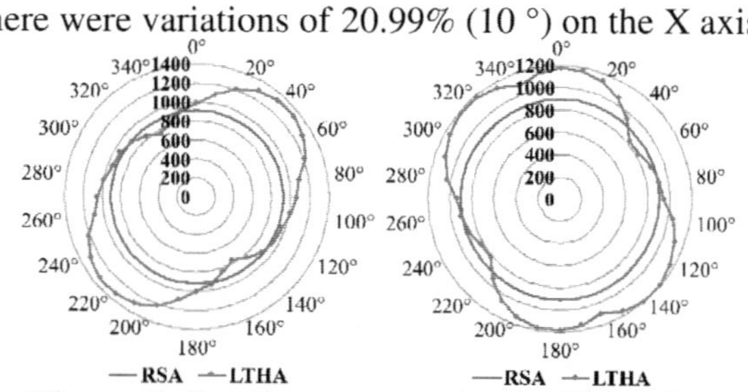

Figure 8. Bending moment (t-m) of Model 1 (X-axis, left and Y-axis, right).

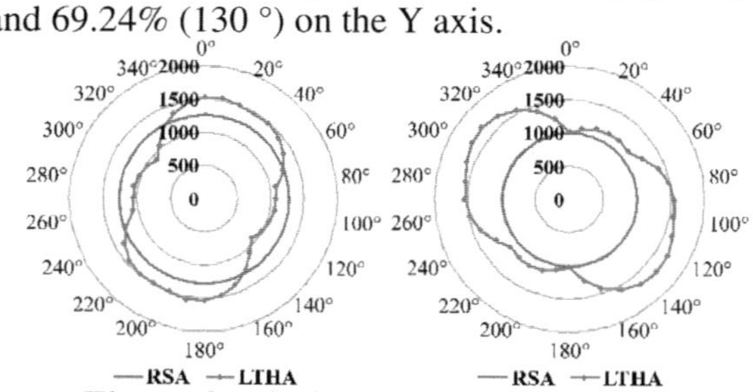

Figure 9. Bending moment (t-m) of Model 2 (X-axis, left and Y-axis, right).

3.5. Lateral displacement

Following figures show lateral displacements in the first level that are taken for each structure analysed. In model 1 (figure 10), variations of up to 31.13% (0 °) can be observed in the X axis and 48.59% (40 °) in the Y axis. On the other hand, in the model 2 (figure 11) are presented variations of up to 71.52% (110 °) on the X axis and 18.95% (190 °) on the Y axis.

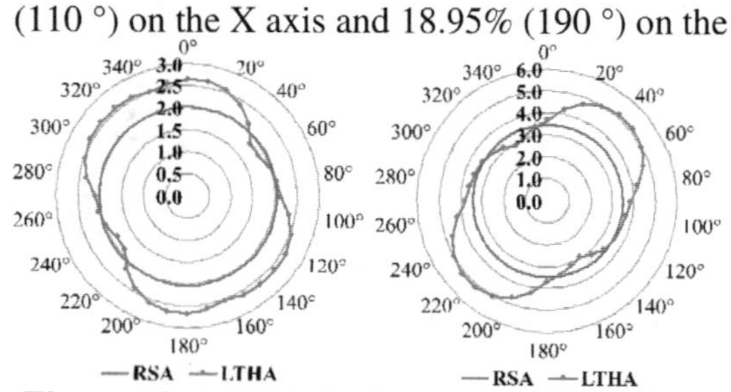

Figure 10. Lateral displacements of first floor (mm) of Model 1 (X-axis, left and Y-axis, right).

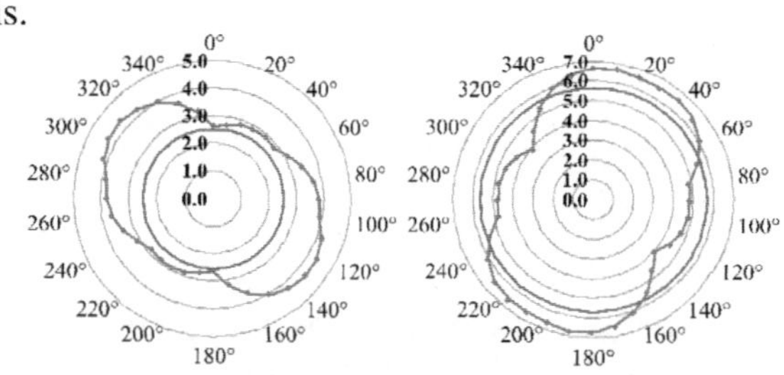

Figure 11. Lateral displacements of first floor (mm) of Model 2 (X-axis, left and Y-axis, right).

With displacement results obtained for each level, drift control has been carried out in accordance with FEMA 356 considerations for reinforced concrete structures [11].

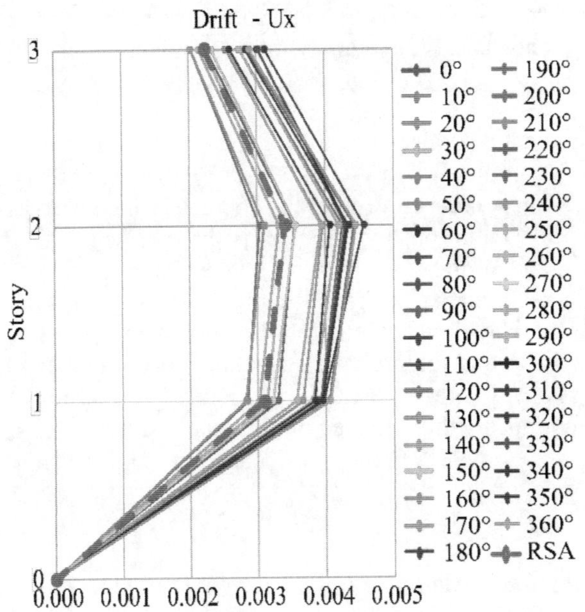

Figure 12. Model 1 drifts, X-axis.

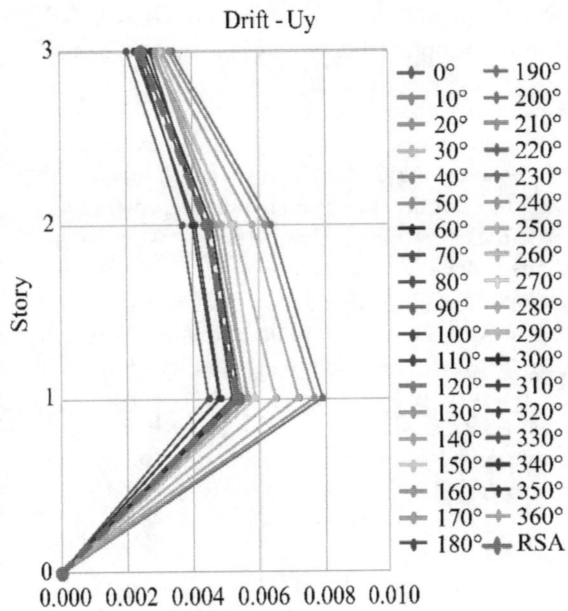

Figure 13. Model 1 drifts, Y-axis.

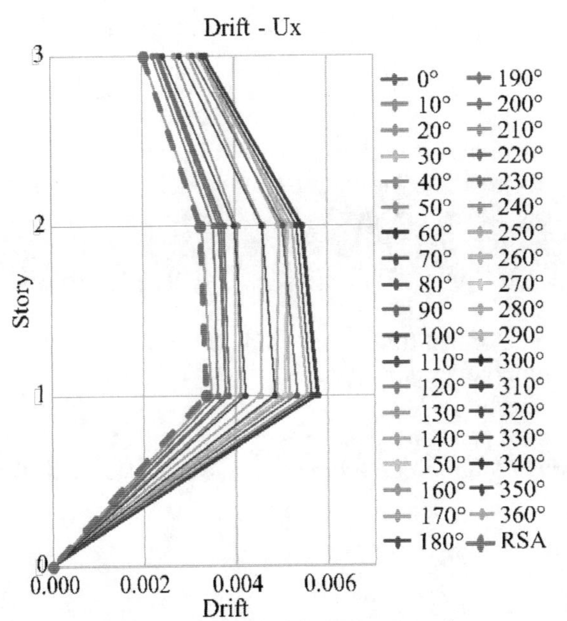

Figure 14. Model 2 drifts, X-axis.

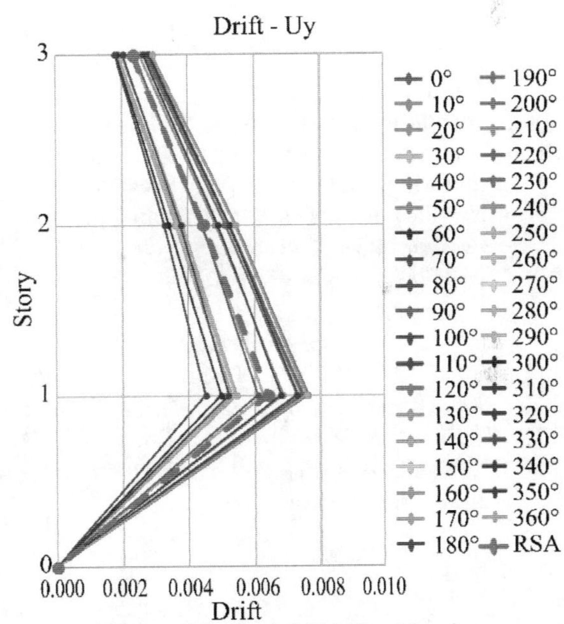

Figure 15. Model 2 drifts, Y-axis.

In the figure 12, figure 13, figure 14 and figure 15 can be observed that in most of the incidence angles analysed the obtained drifts exceed the mezzanine distortion limit obtained according to the RSA. The case of the X resistant axes for both models these don't exceed the seismic performance limit (0.007), since they have been conservatively designed, but in the Y-resistant axes for the two models in some angles of incidence they don't comply with said condition.

3.6. Seismic bidirectionality amplification factor

According to results obtained, the following amplification factors by seismic bidirectionality are considered considering the maximum variations obtained for the studied responses. For structures with irregular incoming corner the factor K is 1.5 and for structures with irregular soft floor K is 1.7. The factor K is applicable for responses such as displacements, basal shear and bending moment.

4. Conclusions

The unidirectional analysis doesn't reflect the most exact responses of buildings, since the maximum responses obtained aren't presented in the axes that are commonly analysed; that is, responses calculated with the unidirectional analysis underestimate actual responses of structures. The effect increases as structures present a greater degree of irregularity within their configuration, as is the case with the l-shaped plane compared to the soft story model, in which variations of up to 50% and 72% respectively are obtained. In view of this, it's necessary to include a factor for increasing global responses in order to consider the seismic bidirectionality in responses calculated through the unidirectional analysis. For structures with l-shaped plan irregularity the factor K is 1.5 and for structures with soft story irregularity the factor K is 1.7. These factors by seismic bidirectionality are applicable to structures with types of irregularity studied. For future investigations it's recommended to analyse buildings considering mass irregularity or vertical discontinuity in structural elements taking into account the seismic bidirectionality.

5. References

[1] Villar-Vega M and Silva V 2017 *Soil Dynamics and Earthquake Engineering* **99** 86-96
[2] Pinzón L, Pujades L, Diaz S and Alva R 2018 *Bulletin of the Seismological Society of America* 5A **108** 2543-55
[3] Nievas C and Sullivan T 2017 *Earthquake Eng. & Struct. Dynamics* **46(5)** 791-809
[4] Kostinakis K, Manoukas G and Athanatopoulou A 2018 *KSCE Journal of Civil Engineering* **22(2)** 725-35
[5] Magliulo G, Maddaloni G and Petrone C 2014 *Earthquake Engineering and Engineering Vibration* **13(2)** 243-56
[6] Cantagallo C, Camata G and Spacone E 2015 *Earthquakes and Structures* **8(1)** 185-204
[7] Amarloo N and Emami A 2019 *Bulletin of Earthquake Engineering* **17** 3447-74
[8] Heredia-Zavoni E and Machicao-Barrionuevo R 2004 *Earthquake Engineering and Structural Dynamics* **33(2)** 271-284
[9] 2019 *Reglamento Nacional de Edificaciones* E030 (Lima: Ministerio de Vivienda Construcción y Saneamiento)
[10] 2012 *Diseño Sísmico de Edificios NCh433* (Santiago: Instituto Nacional de Normalización)
[11] 2009 *FEMA 356* (D.C. United States: Federal Emergency Management Agency)
[12] Paz M 1992 *Dinámica Estructural* (Barcelona: Reverté)
[13] Chopra A 2014 *Dinámica de Estructuras* (México: Pearson Educación)
[14] Hudson D 1990 *Reading and Interpreting Strong Motion Accelerograms* (California: Earthquake Engineering Research Institute)

IOP Conf. Series: Materials Science and Engineering **910** (2020) 012002 doi:10.1088/1757-899X/910/1/012002

Shear capacity of reinforced concrete deep beams using genetic algorithm

Tadesse Wakjira[1]*, Mohamed Ibrahim[1], Bilal Sajjad[1], Usama Ebead[1]

[1]Qatar University, College of Engineering, Civil and Architectural Engineering Department

Corresponding author's e-mail: twakjira@qu.edu.qa

Abstract. It is vital to understand the shear behaviour of reinforced concrete (RC) beams in order to avoid a catastrophic shear failure and design for ductile failure. However, due to the complexity in the shear failure mechanism and various parameters influencing the shear behaviour of RC beams, the accuracy in the determination of the shear capacity remains a challenge. In this paper, machine learning and genetic algorithm are utilized to develop an improved shear design equation for RC deep beams without stirrups. The proposed model considers the parameters influencing the shear capacity of beams including concrete compressive strength, cross-sectional dimension of the beams, aspect ratio, and internal reinforcement ratio. The prediction capability of the proposed model has been compared with that of ACI 318 and resulted in a better prediction in terms of safety, accuracy, and economic aspects.

1. Introduction

An extensive research has been carried out to understand the shear behaviour of reinforced concrete (RC) beams. However, due to the complexity in shear failure mechanism and various parameters influencing the shear capacity, its accurate prediction remains a challenge. Several research contributions have been reported on investigating the shear behaviour of RC beams. The modified compression field theory has shown to be an effective and accurate method for predicting the shear capacity of both pristine [1,2] and strengthened beams [3–8]. Recently, Wakjira and Ebead [9] proposed an improved shear design equation for both pristine and fabric reinforced cementitious matrix (FRCM) strengthened slender beams, considering the important parameters.

Beams with the shear span-to-effective depth (a/d) ratio of less than 2.5 are regarded as deep beams according to ACI-ASCE Committee 445 [10]. Several experimental results have been reported on the shear behaviour of RC deep beams [11–14]. The strut and tie model (STM) is generally used for shear capacity determination of RC deep beams and it has been incorporated in international codes and provisions including ACI 318 [15]. The STM is complicated to be used by design practitioners. Thus, the current study deals with investigating the parameters influencing the shear capacity of RC deep beams internally unreinforced with stirrups and proposing an improved and simple shear design equation for such beams based on artificial intelligence and genetic algorithm approaches. The prediction capability of the proposed equation is compared with that of ACI 318.

2. Methodology

2.1. Database

A database of shear-critical RC deep beams that failed in shear prior to flexural failure have been considered in this study [16]. Table 1 presents the statistical characteristics of shear design parameters for the beams considered in this study. As listed in this table, the database contained a wide range of the design parameters. These parameters include compressive strength of concrete ($f_c{}'$), longitudinal reinforcement ratio (ρ_{sx}), yield strength of longitudinal reinforcement (f_{sx}), cross-sectional width of the beam (b_w), effective depth of the beam (d), and shear span-to-depth (a/d) ratio. The scatter plots of the shear capacity versus each parameter is shown in Figures 1a–1f.

Table 1. Statistical characteristics of the test parameters of the beams

Parameters	Deep beams without stirrups (371 beams)			
	Minimum	Maximum	Mean	Coefficient of Variation
a/d	0.2500	2.4900	1.6374	33.92
$f_c{}'$ (MPa)	20.000	103.400	35.166	47.88
b_w (mm)	76.00	500.00	180.46	38.89
d (mm)	132.00	1097.00	361.81	50.86
ρ_{sx} (%)	0.3400	9.2700	2.3680	57.50
f_{sx} (MPa)	267.00	1480.00	470.04	27.01

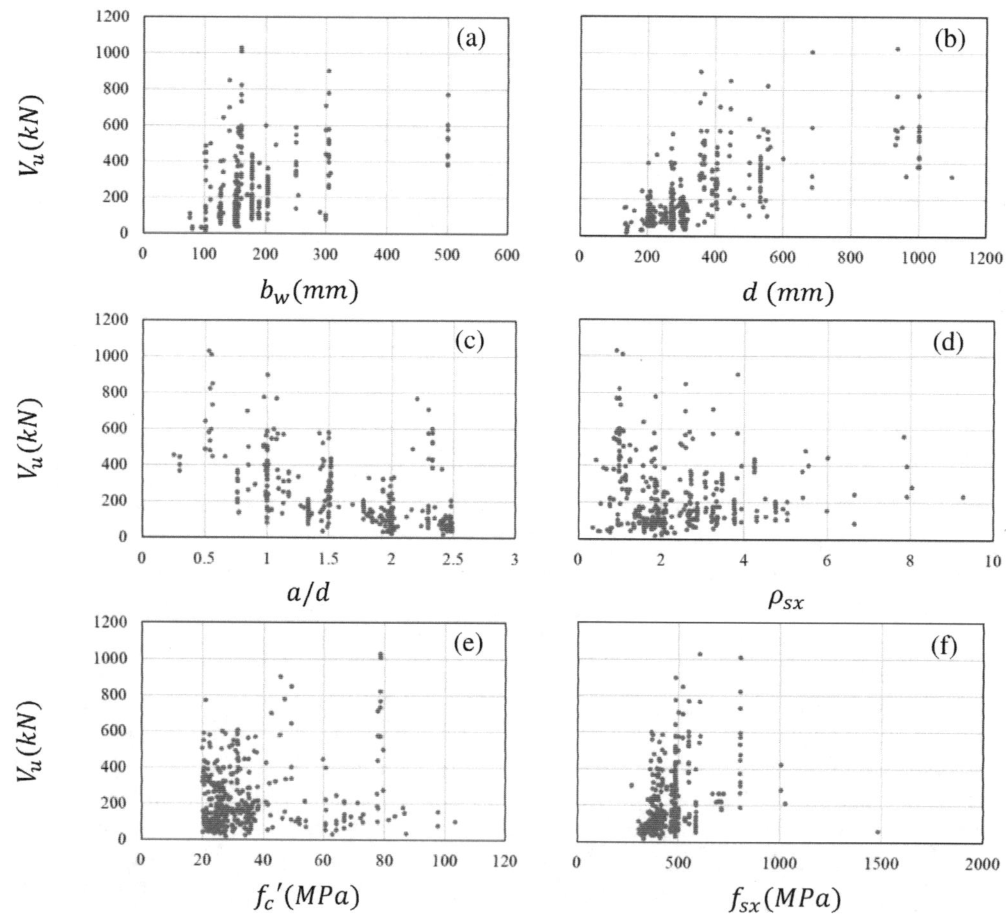

Figure 1. Variation of shear capacity with respect to the design parameters.

2.2. Shear design equation

Machine learning using genetic algorithm (GA) has gained popularity in solving complex nonlinear problems. The GA is a global search procedure for gradually improving the solution in succeeding populations using genetic operators, namely, mutation, selection, and crossover as inspired by Darwin's theory of evolution and the natural law of survival of the fittest [17]. Figure 2 shows the basic steps of the GA.

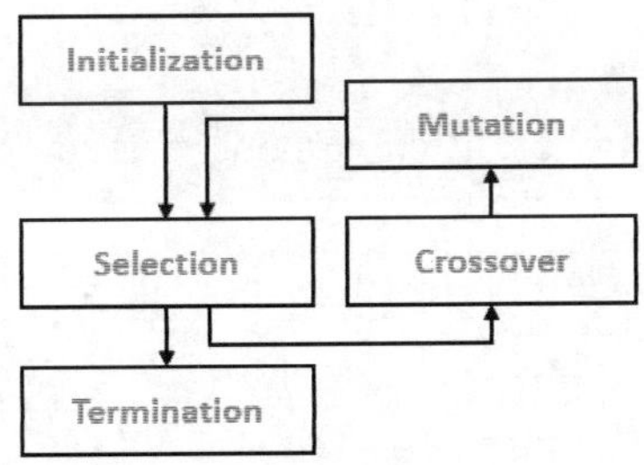

Figure 2. Variation of shear capacity with respect to the design parameters.

Explaining the fundamental basis of GAs is beyond the scope of this paper and details on how to build a genetic algorithm model for optimization can be found elsewhere Goldberg 1989 [18]. In this study, the GA algorithms approach is used herein as an optimization technique to develop equation for the shear capacity of RC deep beams. Thus, an original form of the equation defining the overall shear behaviour with the significant parameters that influence the shear capacity of concrete beams is required [19]. The objective function for the shear capacity of RC beams is defined in terms of the independent variables as follows:

$$V_u = c_1 f_c{}^{c_2}(\rho_{sx} f_{sx})^{c_3}(a/d)^{c_4} b_w d \qquad (1)$$

where, c_1, c_2, c_3, and c_4 are the equation unknown coefficients that are to be estimated.

The GA technique is utilized to develop the shear equation based on the database of 371 shear-critical RC deep beams. From the database, 50% was selected randomly and used for training the model, while the remaining 50% was used for validating the model. The result of the GA model is given in equation below.

$$V_{th} = 0.0456 f_c{}^{0.619} \rho_{sx}{}^{0.411}(a/d)^{-0.874} b_w d \qquad (2)$$

Figures 3a and 3b show the correlation between the predicted and experimental shear capacities based on the proposed equation and ACI 318, respectively. The exact match between the test results and predicted values for shear capacity is shown using the 45° line.

2.3. Assessment of the predictive performance

The prediction of the proposed equation is compared with that of ACI 318-14 [15] in terms of safety, accuracy, and economic aspects based on the modified version of the demerit points classification (DPC) [20] proposed by Collins [21]. In this approach, a penalty is assigned to each value of V_{th}/V_{ex} ratio according to the criteria presented in Table 2 and the total penalty of the model indicates its performance.

Table 2. Demerit Points Classification [20]

Range	Classification	Penalty
$V_{th}/V_{ex} \geq 2$	Extra dangerous	10
$1.176 \leq V_{th}/V_{ex} < 2$	Dangerous	5
$0.869 \leq V_{th}/V_{ex} < 1.176$	Appropriate safety	0
$0.5 \leq V_{th}/V_{ex} < 0.869$	Conservative	1
$V_{th}/V_{ex} \leq 0.5$	Extra conservative	2

Table 3 as well as Figures 3a and 3b show the predictive performance of the proposed model and ACI 318-14 [15]. As can be seen in this figure, the proposed design equation using the GA model provides better predictions than that of ACI 318 with an average V_{th}/V_{ex} of 0.82 and standard deviation (STD) of 0.25. The ACI 318 provision resulted in a lower average of the V_{th}/V_{ex} ratio of 0.79 and higher STD of 0.27, as listed in Table 3. Moreover, the proposed equation resulted in a lower total penalty of 390 compared to that for the ACI 318, which was 403, as listed in Table 3. Thus, it can be concluded that the proposed equation resulted in a better prediction capability compared to that of ACI 318.

Table 3. Prediction capability of different models based on the modified DPC [20].

Model	Total #	V_{th}/V_{ex} Average	V_{th}/V_{ex} STD	Extra dangerous	Dangerous	Appropriate safety	Conservative	Extra conservative	Total Penalty
Proposed model	371	0.82	0.25	–	27	113	207	24	390
ACI 318-14	371	0.79	0.27	–	9	70	226	66	403

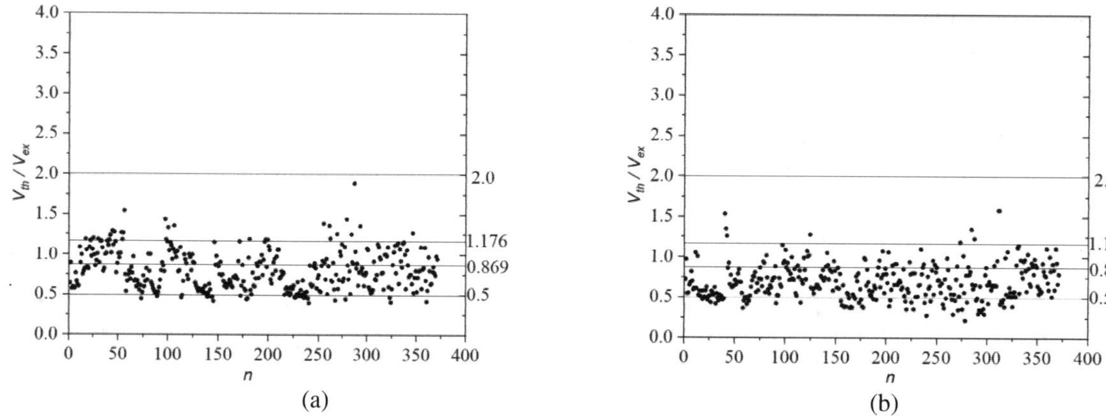

Figure 3. Prediction capability of the proposed equation (a) and ACI 318 (b) based on the DPC [20].

3. Conclusion

In this paper, an improved shear design equation for RC deep beams is proposed based on genetic algorithm. Total of 371 shear-critical RC deep beams was used in the database. Half of the total database has been used for training the GA model, while the remaining was used for validation. The study considered the important parameters that influence the shear capacity of the RC deep beams including compressive strength of concrete, longitudinal reinforcement ratio, yield strength of longitudinal reinforcement, cross-sectional width of the beam, effective depth of the beam, and shear span-to-depth ratio. The prediction performance of the proposed model has been compared to the existing ACI 318-14 provision and resulted in a better prediction capability in terms of safety, accuracy, and economic aspects.

4. References

[1] Bentz EC, Vecchio FJ, Collins MP. Simplified Compression Field Theory for Calculating Shear Strength of Reinforced Concrete Elements. ACI Struct J 2006;103:614–24. https://doi.org/10.14359/16438.
[2] Vecchio FJ, Collins MP. The modified compression-field theory for reinforced concrete elements subjected to shear. ACI J Proc 1986;83. https://doi.org/10.14359/10416.
[3] Ibrahim M, Wakjira T, Ebead U. Shear strengthening of reinforced concrete deep beams using near-surface mounted hybrid carbon/glass fibre reinforced polymer strips. Eng Struct

2020;210:110412. https://doi.org/10.1016/j.engstruct.2020.110412.
[4] Wakjira T, Ebead U. A shear design model for RC beams strengthened with fabric reinforced cementitious matrix. Eng Struct 2019;200. https://doi.org/10.1016/j.engstruct.2019.109698.
[5] Wakjira T, Ebead U. FRCM/internal transverse shear reinforcement interaction in shear strengthened RC beams. Compos Struct 2018;201:326–39. https://doi.org/10.1016/j.compstruct.2018.06.034.
[6] Wakjira T, Ebead U. Shear span-to-depth ratio effect on steel reinforced grout strengthened reinforced concrete beams. Eng Struct 2020;216. https://doi.org/10.1016/j.engstruct.2020.110737.
[7] Wakjira T, Ebead U. A new approach for predicting the shear capacity of FRCM strengthened RC beams in shear. IOP Conf. Ser. Mater. Sci. Eng., vol. 431, 2018. https://doi.org/10.1088/1757-899X/431/11/112006.
[8] Wakjira T, Ebead U. Experimental and analytical study on strengthening of reinforced concrete T-beams in shear using steel reinforced grout (SRG). Compos Part B Eng 2019;177. https://doi.org/10.1016/j.compositesb.2019.107368.
[9] Wakjira T, Ebead U. Simplified Compression Field Theory-Based Model for Shear Strength of Fabric-Reinforced Cementitious Matrix- Strengthened Reinforced Concrete Beams. ACI Struct J 2020;117. https://doi.org/10.14359/51721366.
[10] Joint ACI-ASCE Committee 445. Recent approaches to shear design of structural concrete. J Struct Eng 1998;124:1374–417. https://doi.org/10.1061/(ASCE)0733-9445(1998)124:12(1375).
[11] Siess C. Strength and Behavior of Deep Beams in Shear. J Struct Div 1965;91:19–41.
[12] Quintero-Febres CG, Parra-Montesinos G, Wight JK. Strength of struts in deep concrete members designed using strut-and-tie method. ACI Struct J 2006;103:577–86. https://doi.org/10.14359/16434.
[13] Smith KN, Vantsiotis AS. Shear Strength of Deep Beams. J Am Concr Inst 1982;79:201–13. https://doi.org/10.14359/10899.
[14] Wakjira T, Ebead U. Internal transverse reinforcement configuration effect of EB/NSE-FRCM shear strengthening of RC deep beams. Compos Part B Eng 2019;166:758–72. https://doi.org/10.1016/j.compositesb.2019.03.004.
[15] ACI Committee 318. Building code requirements for structural concrete (ACI 318-14). American Concrete Institute, Farmington Hills, MI, USA.; 2014.
[16] Collins MP, Bentz EC, Sherwood EG. Where is shear reinforcement required? Review of research results and design procedures. ACI Struct J 2008;105:590–600. https://doi.org/10.14359/19942.
[17] Xu B, Lin H, Wagholikar KB, Yang Z, Liu H. Identifying protein complexes with fuzzy machine learning model. Proteome Sci 2013. https://doi.org/10.1186/1477-5956-11-S1-S21.
[18] Goldberg DE. Genetic algorithms in search. Optim Mach 1989.
[19] Nehdi M, El Chabib H, Aly Saïd A. Proposed shear design equations for FRP-reinforced concrete beams based on genetic algorithms approach. J Mater Civ Eng 2007. https://doi.org/10.1061/(ASCE)0899-1561(2007)19:12(1033).
[20] Neto BNM, Barros JAO, Melo GSSA. Model to Simulate the Contribution of Fiber Reinforcement for the Punching Resistance of RC Slabs. J Mater Civ Eng 2014;26:04014020. https://doi.org/10.1061/(ASCE)MT.1943-5533.
[21] Collins MP. Evaluation of shear design procedures for concrete structures. Rep. Prepared for the CSA Technical Committee on Reinforced Concrete Design, Canada: 2001.

Acknoledgement
The authors would like to acknowledge Qatar National Research Fund (a member of Qatar Foundation) for NPRP grant # NPRP 9-110-2-052 and UREP grant # UREP24-045-2-013. The authors would also like to acknowledge Qatar University for the Graduate Assistantship, students' codes GTRA-CENG-2019-09 and GTRA-CENG-2019-16. The findings achieved herein are solely the responsibility of the authors.

IOP Conf. Series: Materials Science and Engineering **910** (2020) 012003 doi:10.1088/1757-899X/910/1/012003

Analysis of Covering Problem Models for Setting the Location of a Ready-Mixed Concrete Plant: Case Study of the Rayong Province, Thailand

Satawat Doungpan

School of Management Science, Sukhothai Thammathirat Open University Nonthaburi 11120, Thailand

Email: kache326@hotmail.com

Abstract. Decision makers must prepare good plans when developing large-scale projects. The Thai government has been increasing economic growth in three provinces under the Eastern Economic Corridor Development Program (EEC). The project includes reconstruction and new construction. This paper aims to find locations for several ready-mixed concrete plants to cover all ready-mixed concrete needs in the Rayong province. The plant location problem could be solved using optimization modelling for a set covering location problem (SCLP) and a maximal covering location problem (MCLP). The results show that two ready-mixed concrete plants location in the Mueang Rayong district and Klaeng district could meet the problem solution.

1. Introduction

Many construction projects have been initiated to support the Thai government's Eastern Economic Corridor Development Project (EEC). New infrastructure and renovation of existing facilities including highways, high-speed trains, an industrial city and an airport have been planned on Chachoengsao, Chonburi, and Rayong, as shown in Figure 1.

The private sector will benefit from these projects by supporting the demand for construction materials. Ready-mixed concrete is an essential construction material. However, supplying ready-mixed concrete is challenging because it takes a significant investment, and other issues must be addressed [1],[2]. Furthermore, the life of ready-mixed concrete is short, and it must be delivered within approximately two hours once it is produced. Delivery time is dependent on distance, driving time and geography. The existing facility location is inappropriate for service due to potential problems in product distribution. As a result, concrete manufacturers cannot provide satisfactory services to all users. The target of this research is:

1. To find the optimum location for the problems in providing ready-mixed concrete service by analyzing the two methods.
2. To compare the result of methods between the set covering location problem (SCLP) and the maximal covering location problem (MCLP)
3. To compare the advantage and disadvantages of the two methods.

This paper is organized as follows: In Section 2, we review the literature addressing covering problem models for each field. In Section 3, the methodology of the two models is presented. Section 4 describes vital data and assumptions of the case study. In Section 5, the result is analyzed and discussed. Finally, the last section provides a conclusion.

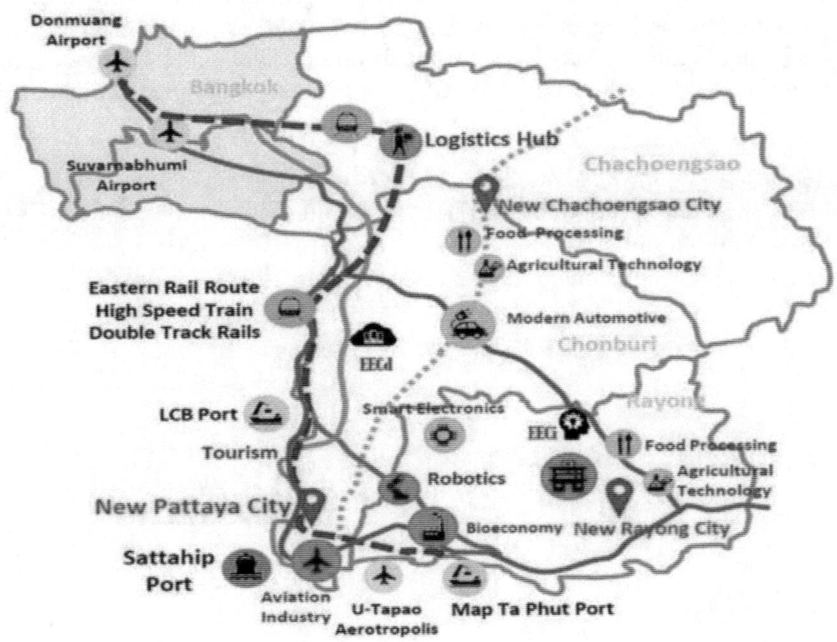

Figure 1. An overview of the EEC. Source: The Eastern Economic Corridor Office of Thailand

2. Literature Review

The literature related to the covering problems in this paper can be divided into two main topics: SCLP and MCLP, which were formulated as the integer programming problem (IP) by [3], [4] and [5]. By now, The SCLP and MCLP are the most extensive and classical models of finding the optimal solution for a decision problem in which the decision variable is binary 1 or 0 (open or closed). Many researchers [6] studied and adopted the covering problem in various research areas. However, we found that there were few studies on the construction management field [7], especially for applying the covering problem models with real cases. We summarize the literature review in Table 1.

Table1. Summary of the covering problem for an application and extension for each the research areas

Problems Type		Applications	References
SCLP	MCLP		
📍		Ready-Mix Concrete Plant	[7]
📍		Ambulance location	[8]
📍		Emergency facility location	[9]
📍		Bus stops location	[10]
📍	📍	Healthcare facility location	[11]
	📍	Ambulance location	[12]
	📍	The workload capacity for a facility	[13]
	📍	Emergency Service Facilities	[14]
	📍	Hub location	[15]

3. Methodology

The objective of this work is to find the optimal solution of setting up ready-mix concrete plant sites to serve the new construction projects in Rayong province. To clarify the research results, we experimented by applying both the SCLP model and the MCLP model to the problem and compared the results in

IOP Conf. Series: Materials Science and Engineering **910** (2020) 012003 doi:10.1088/1757-899X/910/1/012003

many dimensions. Because the target of the study aims to find the minimum number of plants required, the integer programming for both covering problem models could guarantee that every demand node is covered by at least one facility within the determined time (the maximum delivery time constraint). Therefore, the solution could be achieved. We could formulate the mathematical models for SCLP and MCLP as follows:

3.1. The set covering location problem (SCLP)

The SCLP was first proposed originally by [4], and the solution provided by this method includes an unlimited number of facilities unless specified otherwise. SCLP is an appropriate model that can cover all users and low construction costs due to the objective function to minimise.

$$Min\ z = \sum_{j=1}^{n} c_j x_j \tag{1}$$

Subject to

$$\sum_{j=N_i}^{n} t_{ij} x_j \geq 1 \quad \forall i \in I \tag{2}$$

$$x_j \in \{1,0\} \quad \forall j \in J \tag{3}$$

Objective functions and constraint functions:

Equation (1) the objective function (z) desires the minimum number of facility locations j (a ready-mixed concrete plant) that cover all demand nodes i under the maximum delivery time. Equation (2) defines the constraint that each demand node must be served by at least one facility. Equation (3) defines the decision variable x_j as binary variable 1 or 0.

3.2. The maximal covering location problem (MCLP)

The MCLP was first presented originally by [5], so to solve the problem with the smallest number of facilities. This problem agrees with real situations in the construction industry better than the SCLP when the manager must consider the construction cost in the planning phase.

$$Max\ z = \sum_{j=1}^{n} w_i y_i \tag{4}$$

Subject to

$$\sum_{j=N_i}^{n} t_{ij} x_j \geq y_i \quad \forall i \in I \tag{5}$$

$$\sum_{j=1}^{n} x_j = P \tag{6}$$

$$x_j \in \{1,0\} \quad \forall j \in J \tag{7}$$

$$y_i \in \{1,0\} \quad \forall i \in I \tag{8}$$

Objective functions and constraint functions:

In Equation (4) the objective function (z) is to maximize the demand nodes coverage, under the maximum delivery time that can be covered with the number of facilities. In Equation (5), the constraint defines that all demand nodes must be serviced from facility locations within the coverage travel time from at least one facility that was set. In Equation (6), the constraint defines that the least coverage (at least one a possible facility, it must previously set). Equation (7) and (8) are the constraint defines a decision variable x_j and y_i as binary variable 1 or 0 All the sets and indices the parameters and the decision variables of the SCLP and MCLP are shown in Table 2.

We explain that the different points of mathematical equations between the two methods are:

1. The MCLP adds a binary decision variable for considering the facility that covers the demand nodes y_i, but the SCLP does not consider the amount of supply required by each demand node.
2. The MCLP can add the parameter P for limiting the number of facilities, but the SCLP cannot limit the number of facilities.

Table 2. The Description of an element of two methods for generating models

Symbol	Description
Sets and Indices:	
$i \in I$	The set of demand nodes supersede with the districts into Rayong province.
$j \in J$	The set of facility locations available to be selected for setting a ready-mixed concrete plant
Parameters:	
c_j	The construction cost of a facility (ready-mixed concrete plant) at location j
t_{ij}	The traveling time between the demand nodes i and possible facility locations j
N_i	$\{j \mid t_{ij} \leq S\}$ The set of all possible facility locations j that can cover the demand nodes i within the maximum delivery time constraint.
S	The maximum delivery time to ensure the service quality according to the pragmatic standard.
P	The number of facilities which are previously assigned.
w_i	The demand node at i (The average amount of ready-mixed concrete (m^3/day))
Decision Variables:	
x_j	A binary variable for setting the location j as a ready-mixed concrete plant site.
	$x_j = 1$, if the facility j is chosen to be setting up a ready- mixed concrete plant, $x_j = 0$, otherwise
y_i	A binary decision variable indicates that the facility can covered the demand nodes at i
	$y_i = 1$, if the demand nodes at i covered , $y_i = 0$, otherwise

4. Case study

Rayong province is in the eastern part of Thailand and borders the Gulf of Thailand at the north. There are many heavy industries and much tourism in this area. Rayong was selected for our study because it is one of the provinces in the EEC pilot project, a large-scale, Thai-government project. We conducted field surveys at Rayong, including road networks, shortcuts, and traffic junctions.

4.1. The Road network and Data

All data collected for maps are publicly available and provided by The Ministry of Tourism and Sports Thailand [16]. The travelling time was taken from the geographic information system (GIS) [17]. via the web service provided by the publicly available, The Ministry of Transport, Thailand. These data are synthesised to give an important parameter before feeding the data to both the SCLP and MCLP models, as shown in Table 3: Travel time in the SCLP and MCLP.

4.2. Comprehensive assumption of models

The SCLP model can solve the covering location problem and produces the ideal number of facilities without any limiting constraint. However, this solution may not be practical for real business where investment policies could be applied as constraints to the model. Thus, the MCLP [4] was created to support additional system parameters such as the maximum number of facilities and construction costs of individual facilities. Moreover, the MCLP is capable of searching for new facility locations while the existing facilities are considered. This additional functionality helps demand nodes proliferate to receive more comprehensive services. However, both SCLP and MCLP are deterministic versions, so all assumed parameters must be clearly stated and consistent with the real world. These techniques have been validated against various resource allocation problems, construction site problems, and other interesting issues [18]. The solution can be considered reliable and optimal because our study applied both techniques and found that the solutions produced by SCLP and MCLP aligned.

4.3. Result

Quality control of the concrete-supply service is the most important factor in our research. Therefore, this research applied recommendations of civil engineering experts and the Institute of Engineering of Thailand (EIT) standards [19] to control the quality of construction. We used the maximum delivery time of 60 minutes in the SCLP, and the MCLP models include the same parameters. The results of both SCLP and MCLP are shown in Table 4.

Table 3. Supersedes the travel time of a concrete mixing truck between a pair of a demand node.

No.	District	Mueang Rayong	Ban Chang	Klaeng	Wang Chan	Ban Khai	Pluak Daeng	Khao Chamao	Nikhom Phattana
1	Mueang Rayong	0	20	34	44	11	38	55	21
2	Ban Chang	20	0	50	57	25	36	71	19
3	Klaeng	34	50	0	17	40	57	28	48
4	Wang Chan	44	57	17	0	33	41	22	46
5	Ban Khai	11	25	40	33	0	31	55	14
6	Pluak Daeng	38	36	57	41	31	0	62	18
7	Khao Chamao	55	71	28	22	55	62	0	68
8	Nikhom Phattana	21	19	48	46	14	18	68	0

4.4. Analysis Result

The methods of the SCLP and the MCLP was modelled using software General Algebraic Modeling System (GAMS) and chose XPRESSMP (Trial Version Free Download) The results are analyzed and explained as follows:

1.The SCLP and MCLP provide the same results when the delivery time is 60 minutes. Both methods showed an ideal result of two facilities to supply sufficient ready-mixed concrete for each demand node.

2. We decided to locate the ready-mixed concrete plants at Mueang Rayong and Klaeng district to cover the desired nodes throughout the province with minimum construction cost. These results will be valuable for decision-makers and project managers in planning and construction management.

Table 4. The result of the SCLP and MCLP models.

No.	District	The demand nodes	SCLP	MCLP								
			$S \leq 60$	$S \leq 60$								
			$P=0$	$P=1$			$P=2$			$P=3$		
		w_i	x_j	x_j	y_i	z	x_j	y_i	z	x_j	y_i	z
1	Mueang Rayong	185	1		1	975	1	1	1,100	1	1	1,100
2	Ban Chang	155	0		1			1			1	
3	Klaeng	175	1		1		1	1			1	
4	Wang Chan	125	0	1				1			1	
5	Ban Khai	130	0		1			1			1	

No.	District	The demand nodes	SCLP	MCLP								
			$S \le 60$	$S \le 60$								
			$P = 0$	$P = 1$			$P = 2$			$P = 3$		
		w_i	x_j	x_j	y_i	z	x_j	y_i	z	x_j	y_i	z
6	Pluak Daeng	145	0		1			1		1	1	
7	Khao Chamao	85	0		1			1		1	1	
8	Nikhom Phattana	100	0		1			1			1	
	Min z		2									
	Max z					975			1,100			1,100

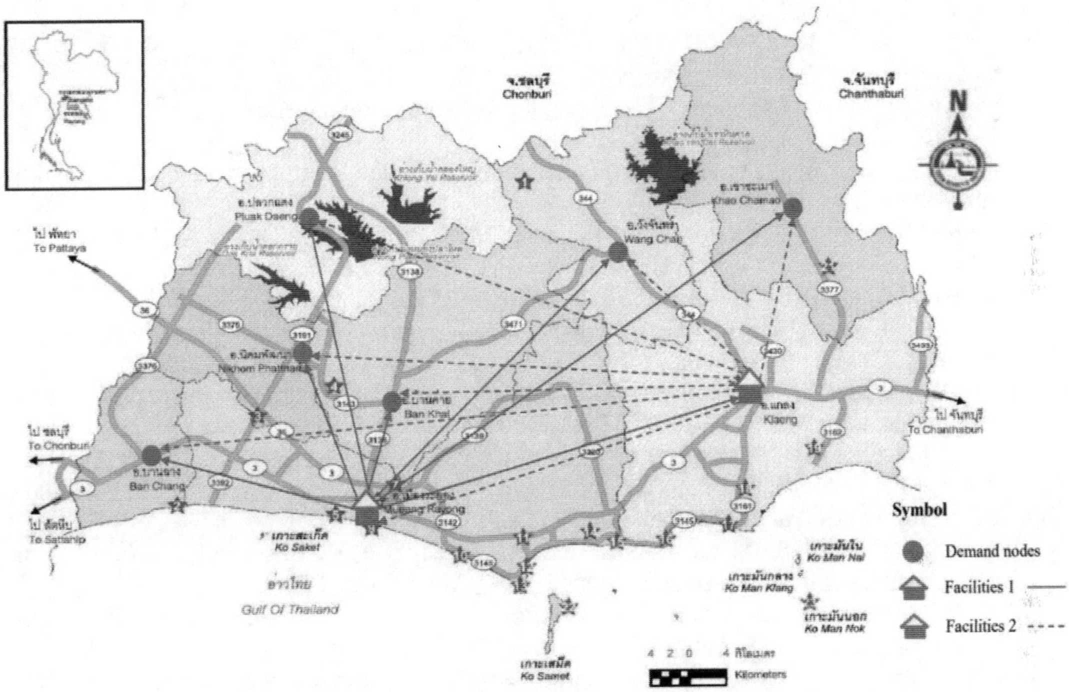

Figure 2. Interactive maps that prove data visualization

5. Conclusion

An optimization technique is applied for finding the optimal solution to a problem of locating production facilities in the optimal location, and then the results are adrenalized.

The main conclusions of this research are as follows:

1. The SCLP and MCLP models can be applied to this situation, following the Engineering Institute of Thailand standards [19] and civil engineering recommendations. These standards and guidelines are designed to uphold the strength and quality control of ready-mixed concrete used for structural purposes.
2. Two facilities are required to cover all the demand nodes of the province. Furthermore, if cheaper costs are desired for constructing a new facility, two facilities are sufficient.
3. The SCLP and MCLP methods were compared, using similar objectives and different constraints. The two methods provided similar results, as shown in Table 4. As a result, the new positions of the ready-mixed concrete plant should be set at the Mueang Rayong and Klaeng districts. See Figure 2 to clarify conclusions 1, 2 and 3
4. The SCLP and MCLP methods had individual advantages and disadvantages.

IOP Conf. Series: Materials Science and Engineering **910** (2020) 012003 doi:10.1088/1757-899X/910/1/012003

a. The MCLP limits the number of facilities, but the SCLP does not. The MCLP can be used to forecast the need for material such as cement sand and rock at each demand node. On the other side, these data can be used to prepare for designing the concrete mixture of the plant (in the overview).

b. The SCLP focuses on finding the minimum number of facilities that can cover the region within the maximum delivery time only. However, the real situation must consider many other dimensions of the problem.

6. References

[1] Balcik B, Beamon BM. Facility location in humanitarian relief. International Journal of Logistics. 2008 Apr 1;11(2):101-21.

[2] Van Wassenhove LN. Humanitarian aid logistics: supply chain management in high gear. Journal of the Operational research Society. 2006 May 1;57(5):475-89.

[3] Toregas C, Revelle C. Binary logic solutions to a class of location problem. Geographical Analysis. 1973 Apr;5(2):145-55.

[4] Toregas C, Swain R, ReVelle C, Bergman L. The location of emergency service facilities. Operations research. 1971 Oct;19(6):1363-73.

[5] Church R, ReVelle C. The maximal covering location problem, Papers Regional Sd. Assoc. 1974, 32(1), 101–118.

[6] Owen SH, Daskin MS. Strategic facility location: A review. European journal of operational research. 1998 Dec 16;111(3):423-47.

[7] Doungpan S. Application the Facility Location Model for Setting Ready-Mix Concrete Plant: Case Study at Rayong Province, Thailand. Proceedings 2020 IEEE 7th International Conference on Industrial Engineering and Applications (ICIEA 2020). Bangkok, Thailand

[8] Brotcorne L, Laporte G, Semet F. Ambulance location and relocation models. European journal of operational research. 2003 Jun 16;147(3):451-63.

[9] Eaton DJ, Daskin MS, Simmons D, Bulloch B, Jansma G. Determining emergency medical service vehicle deployment in Austin, Texas. Interfaces. 1985 Feb;15(1):96-108.

[10] Gleason JM. A set covering approach to bus stop location. Omega. 1975 Oct 1;3(5):605-8.

[11] Fo AR, da Silva Mota I. Optimization models in the location of healthcare facilities: a real case in Brazil. Journal of Applied Operational Research. 2012;4(1):37-50.

[12] Adenso-Diaz B, Rodriguez F. A simple search heuristic for the MCLP: Application to the location of ambulance bases in a rural region. Omega. 1997 Apr 1;25(2):181-7.

[13] Pirkul H, Schilling DA. The maximal covering location problem with capacities on total workload. Management Science. 1991 Feb;37(2):233-48.

[14] Zhang B, Peng J, Li S. Covering location problem of emergency service facilities in an uncertain environment. Applied Mathematical Modelling. 2017 Nov 1;51:429-47.

[15] Peker M, Kara BY. The P-Hub maximal covering problem and extensions for gradual decay functions. Omega. 2015 Jul 1;54:158-72.

[16] The Ministry of Tourism and Sports of Thailand, the publish data [Online]. Available: https://www.tat.or.th/th/central-data/e-brochure?page=4&destination_id=465

[17] The Ministry of Transport, Thailand, via the web service provided by publishing data [Online]. Available: http://www.doh.go.th/content/page/page/8076#

[18] Li H, Love PE. Genetic search for solving construction site-level unequal-area facility layout problems. Automation in construction. 2000 Mar 1;9(2):217-26

[19] The Engineering Institute of Thailand Under H.M. The King's Patronage, EIT Standard 1008-38 and 011014-19

IOP Conf. Series: Materials Science and Engineering **910** (2020) 012004 doi:10.1088/1757-899X/910/1/012004

Performance of Basalt Fiber Reinforced Continuous Beams with Basalt FRP Bars

Abdelrahman Abushanab[1] and Wael Alnahhal[1]

[1] Department of Civil and Architectural Engineering, Qatar University, Qatar.

Email: wael.alnahhal@qu.edu.qa

Abstract. The state of Qatar is suffering from a harsh environment and coastal conditions which stand for most of the year. As a result, steel-reinforced concrete (RC) structures are subjected to rapid corrosion and deterioration. Therefore, there is a necessity to replace the conventional steel reinforcement by fiber-reinforced polymers (FRP) composites. Apart from FRP corrosion resistance property, their strength to weight ratio is higher than steel reinforcement which made them a viable alternative to steel reinforcement. This study aims to investigate the flexural behavior of basalt fiber reinforced concrete continuous beams reinforced with basalt FRP (BFRP) bars. Two RC continuous beams with a size of 200 x 300 x 4000 mm were prepared and cured for 28 days. Then, the beams were tested up to failure under five-point loading. The volume fraction of basalt macro-fibers (BMF) was the only investigated parameter in this study. Test results have shown that the addition of BMF in volume fractions of 0.75% improved the crack widths, cracking loads, failure mode, strains and displacements of the beams.

1. Introduction

Corrosion in steel reinforcement has been identified as one of the most challenging problems in the construction industry. In recent decades, a huge amount of money has been spent to improve the steel reinforcement corrosion resistance in reinforced concrete (RC) structures [1]. However, corrosion was not fully eliminated. For that, acquiring an alternative reinforcement material is needed to overcome the steel corrosion-related problems.

Recently, fiber reinforcement polymers (FRP) composites have been used as an alternative to steel reinforcement to overcome the steel corrosion-related problems. FRP composites have several applications such as internal reinforcements in new RC members, strengthening and retrofitting of existing or failure RC members and chopped fibers in concrete mixtures [2, 3]. Apart from its corrosion resistance, FRP bars have much less weight compared to steel reinforcements [4].

With the development of FRP composites, basalt FRP (BFRP) has been added to the FRP family. BFRP is made by melting the igneous rock at 1400 ^{o}C. On the one hand, it is a non-toxic and environmentally safe material. In addition, it can be available in different forms such as reinforcing bars, mesh and chopped fibers. On the other hand, BFRP has linear-elastic behavior with brittle failure, lower strain at failure and lower modulus of elasticity [2, 5].

Indeterminate RC members such as continuous beams are the most widely used members in real life. A number of studies have investigated the flexural behavior of basalt FRP-RC beams [6-11]. For example, Habeeb and Ashour [11] studied the flexural performance of Glass FRP-RC continuous beams. It was found that the maximum applied load increased with increasing the bottom reinforcement ratio at an increasing rate. It was also observed that crack widths of FRP-RC beams were wider than steel-RC

IOP Conf. Series: Materials Science and Engineering **910** (2020) 012004 doi:10.1088/1757-899X/910/1/012004

beams. Other researchers found that concrete shortcoming properties such as low tensile strength, brittle failure and wider cracks can be mitigated by mixing chopped fibers randomly in concrete mixtures [12-14].

Although many studies investigated the flexural behavior of BFRP-RC simply supported beams and slabs, still further studies are needed to understand the flexural performance of BFRP-RC continuous beams and to establish its design code and guideline. This study was therefore needed to investigate the flexural behavior in terms of cracking loads, strains and displacements of BFRP-RC continuous beams mixed with basalt macro-fibers (BMF).

2. Experimental program

The study consisted of two continuous RC beams measured 4000 mm long and a cross-section of 200 × 300 mm as presented in Figure 1. The first beam was made with plain concrete and the second with basalt fiber reinforced concrete. The beams were loaded until failure under five-point loading using a 1500 kN universal testing machine (UTM). The beams were reinforced with similar longitudinal and transverse reinforcement, but with a different volume fraction of BMF. Volume fraction of BMF of 0.75 was selected in this study based on the findings of a pilot study. Cracking loads, strains and displacements of the tested beams were investigated in this study. The displacement values at each midspan were recorded using a linear variable differential transformer (LVDT) installed at the center of each midspan. The reinforcement details of the test beams are provided in Table 1.

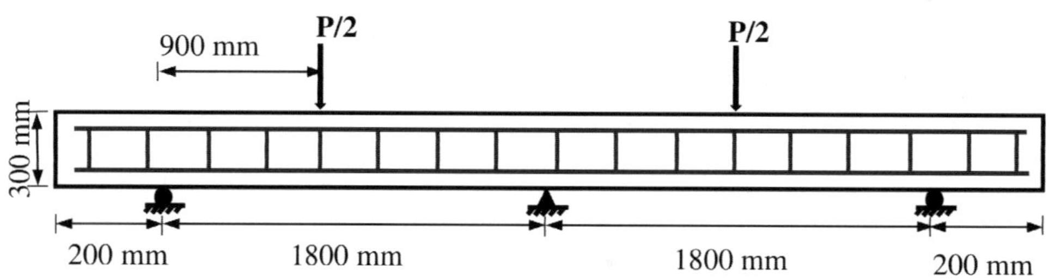

Figure 1. Geometry and details of the tested beams.

Table 1. Reinforcement details of the tested beams.

	Volume fraction of BMF	Bottom reinforcement	Top reinforcement
B1	0.00%	6 ϕ 10	4 ϕ 10
B2	0.75%	6 ϕ 10	4 ϕ 10

3. Results and discussions

3.1. Failure mode and cracking pattern

Table 2 lists the cracking load and the failure mode of beams B1 and B2. It was noticed in beam B1 that as the load increased, the cracks were propagated toward the compression side and became wider until the failure in beam B1. As a result, these cracks caused beam B1 to fail in shear failure mode. By contrast, in beam B2, these cracks were restrained when the volume fraction of BMF of 0.75% was added. The added BMF had prevented the shear failure and caused it to fail in concrete crushing mode. Additionally, it was noticed that the cracking load has increased by 10% in beam B2. The change in the failure mode and the increase in the cracking load are attributed to the bridging effect of the BMF, which had reduced the crack widths. The findings are matching with the findings of Alnahhal and Aljidda [14], who showed that RC beams with BMF experienced less number of cracks.

Table 2. Failure mode, cracking load and crack width of the tested beams.

	Failure mode	Cracking load (kN)
B1	Shear	80.0
B2	Concrete crushing	89.0

3.2. Load-displacement behavior

Figure 2 illustrates the load-displacement curves of beams B1 and B2. The results revealed that both curves were bilinear regardless of the volume fraction of BMF. Both beams demonstrated small displacements before cracking. However, the displacements were significantly increased after cracking. As expected, beam B2 showed better performance in terms of its load-displacement behavior compared to beam B1. However, both beams had approximately similar failure load, but the deflection in beam B2 was lower than beam B1. For example, at the load of P = 150 kN, the measured deflections of beams B1 and B2 were 6.80 and 4.80 mm, respectively. The result of Alnahhal and Aljidda [14] also confirmed that adding the BMF in RC beams increased the ultimate flexural capacity of the beams.

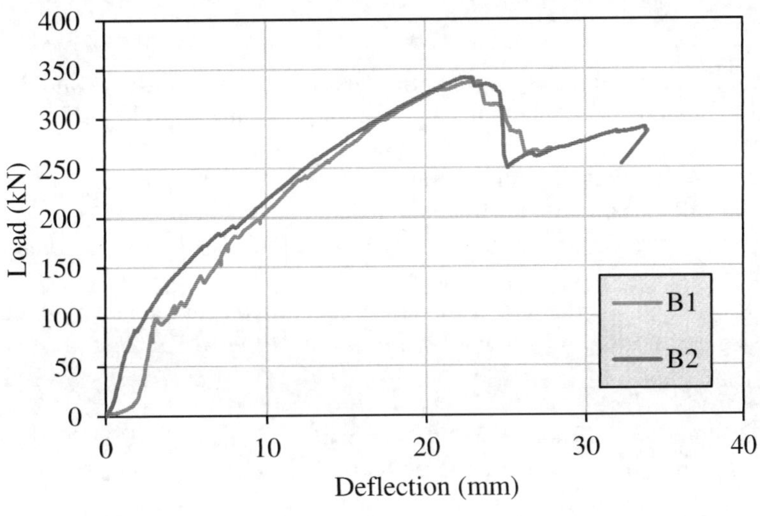

Figure 2. Load-displacement relationship of the tested beams.

3.3. BFRP tensile strain

The variation of the measured strains in BFRP bars between beams B1 and B2 is presented in Figure 3. It is worth noticing that regardless of the BMF, the BFRP tensile strains for both beams increased linearly up to the cracking stage. After that, the strains increased significantly and nonlinearly. At the same load of P = 150 kN, beam B2 had approximately 34% less BFRP strain than beam B1. The reduction in BFRP strains in beam B2 is attributed to the reduced crack widths by the bridging effect as presented in Section 3.2 which, in turn, has controlled the tensile strains in the BFRP bars.

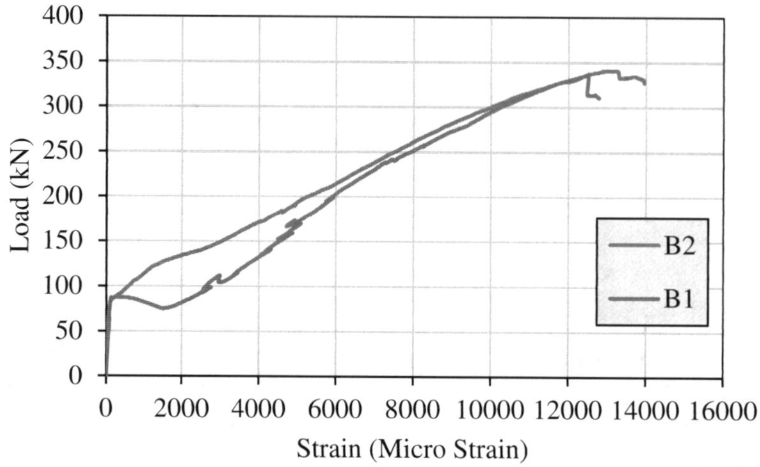

Figure 3. BFRP tensile strain variation against the applied load.

4. Conclusions

This study evaluated the effect of reinforcing continuous RC beams with BFRP and BMF on their failure mode, cracking loads, load-displacement and load-strain relationships. The following conclusions can be drawn from this study:

- The addition of BMF resulted in an improvement in the ultimate flexural capacity, compared with the controlled beam with no BMF.
- The ultimate deflection corresponding to the beam with BMF is lower than the beam with no BMF.
- The cracking pattern and the failure mode were significantly affected by the BMF.
- The BMF significantly improved the tensile strains of the BFRP bars.

5. References

[1] Koch G H, Brongers M P, Thompson N G, Virmani Y P and Payer J H 2002 Corrosion cost and preventive strategies in the United States *Federal Highway Administration (No. FHWA-RD-01-156,, R315-01)* (United States)

[2] High C, Seliem H M, El-Safty A and Rizkalla S H 2015 Use of basalt fibers for concrete structures *Constr. Build. Mater.* **96** 37-46

[3] Adam M A, Said M, Mahmoud A A and Shanour A S 2015 Analytical and experimental flexural behavior of concrete beams reinforced with glass fiber reinforced polymers bars *Constr. Build. Mater.* **84** 354-66

[4] Cai J, Pan J and Zhou X 2017 Flexural behavior of basalt FRP reinforced ECC and concretebeams *Constr. Build. Mater.* **142** 423-30

[5] Elgabbas F, Ahmed E A and Benmokrane B 2015 Physical and mechanical characteristics of new basalt-FRP bars for reinforcing concrete structures *Constr. Build. Mater.* **95** 623-35

[6] Rahman S H, Mahmoud K and El-Salakawy E 2016 Behavior of glass fiber–reinforced polymer reinforced concrete continuous T-beams *J. Compos. Constr.* **21** 04016085

[7] El-Mogy M, El-Ragaby A and El-Salakawy E 2010 Flexural behavior of continuous FRP-reinforced concrete beams *J. Compos. Constr.* **14** 669-80

[8] Attia K, Alnahhal W, Elrefai A and Rihan Y 2019 Flexural behavior of basalt fiber-reinforced concrete beams reinforced with BFRP and GFRP bars *Compos. Struct.* **211** 1-12

[9] El-Mogy M, El-Ragaby A and El-Salakawy E 2011 Effect of transverse reinforcement on theflexural behavior of continuous concrete beams reinforced with FRP *J. Compos. Constr.* **15** 672-81

[10] Akiel M S, El-Maaddawy T and El Refai A 2018 Serviceability and moment redistribution of continuous concrete members reinforced with hybrid steel-BFRP bars *Constr. Build. Mater.* **175** 672-81

[11] Habeeb M N and Ashour A F 2008 Flexural behavior of continuous GFRP reinforced concrete beams *J. Compos. Constr.* **12** 115-24

[12] Ayub T, Shafiq N and Nuruddin M F 2014 Effect of chopped basalt fibers on the mechanical properties and microstructure of high performance fiber reinforced concrete *Adv. Mater. Sci. Eng.* **2014** 187–93

[13] Alnahhal W, Taha R, Alnuaimi N and Al-Hamrani A 2018 Properties of fibre-reinforced concrete made with discarded materials *Mag. Concr. Res.* **71** 152-62

[14] Alnahhal W and Aljidda O 2018 Flexural behavior of basalt fiber reinforced concrete beams with recycled concrete coarse aggregates *Constr. Build. Mater.* **169** 165-78

Acknowledgement
The authors wish to thank Qatar University and Qatar National Research Fund (QNRF) under Qatar Research Leadership Program - Graduate Sponsorship Research award "GSRA6-1-0509-19022" for their financial support.

IOP Conf. Series: Materials Science and Engineering **910** (2020) 012005 doi:10.1088/1757-899X/910/1/012005

CBR behavior of sandy soil reinforced by geofiber material

Omer Mughieda

Abu Dhabi University, Department of Civil Engineering, Abu Dhabi, UAE

Email: omer.mughieda@adu.ac.ae

Abstract. Abu Dhabi sand lacks the required engineering properties to suffice as pavement base course, subbase course, subgrade and as a foundation-support layer under buildings and various structures. Low bearing capacity is manifested on the roadways as differential settlement, waviness and undulations. Stabilization of this type of sand is one of the major geotechnical challenges in practice. In this study the sandy soil was improved by mixing with geofiber. Three different levels of geofiber was added to the sandy soil; 0.1, 0.2 and 0.4% by dry weight of sand. The results showed great increase in shear strength as indicated by the CBR values. The CBR values have increased from 19 to 55 for 0 and 0.4% geofiber content respectively.

1. Introduction

Abu Dhabi sand lacks the required engineering properties to suffice as pavement base course, subbase course, subgrade and as a foundation-support layer under buildings and various structures. The low density of Abu Dhabi sand due to its gradation is mainly resulted in low bearing capacity which is manifested on the roadways as differential settlement, waviness and undulations. This kind of weakness in subgrade causes a lot of distresses in the pavement and cost a lot effort and cost on maintenance and rehabilitation of the pavement. Stabilization of this type of sand is one of the major geotechnical challenges in practice. Improvement of soils with the use of geofiber is popular among practitioners due to low cost and ease of deployment of the technology, light weight of the additive (i.e., geofiber), and successful case histories. The early studies on reinforcement of sand with fibers was performed by Gray and Ohashi [1] how concluded that the shear strength of fiber reinforced sand has increased substantially as well as sand ductility. Using fiber from steel bars, polypropylene, polyester, glass fibers, and biodegraded fibers has proven to be effective for soil reinforcement [2-4]. Polypropylene fiber is the most widely used fibers to enhance the soil strength properties, to reduce the shrinkage properties and to overcome chemical and biological degradation [5-7]. In the current research geofiber with varying content was used to increase the bearing capacity and strength in general of Abu Dhabi sand.

2. Experimental work

2.1. Materials

First, the soil used in this research was Abu Dhabi sand. Basic index properties of the sand were determined according to pertinent ASTM procedures [8,9], and are listed in Table 1. The particle size distribution of the sand is shown in Figure 1. It is classified as poorly graded sand (SP) according to USCS classification. This type of sand is common in UAE in particular and in gulf region in general.

Second, polypropylene geofiber was used as a stabilization material. Many researchers used geofiber in soil strengthening due to its availability, resistance to ultraviolet degradation, chemical stability, and reasonably high strength characteristics [10,11]. The properties of the geofiber used are listed in Table 2. For reinforcing the sandy soil three different levels of geofiber; 0.1, 0.2, and 0.4 % by dry weight of the sand were used in this study.

Table 1. Index properties of the sand investigated

Index properties of Abu Dhabi Sand	
Specific Gravity (ASTM C128) [1]	2.65
D_{10} (mm) (ASTM D6913) [2]	0.08
D_{30} (mm) (ASTM D6913) [2]	0.1
D_{60} (mm) (ASTM D6913) [2]	0.15
C_u (ASTM D6913) [2]	1.875
C_c (ASTM D6913) [2]	0.83
USCS Soil Classification	SP

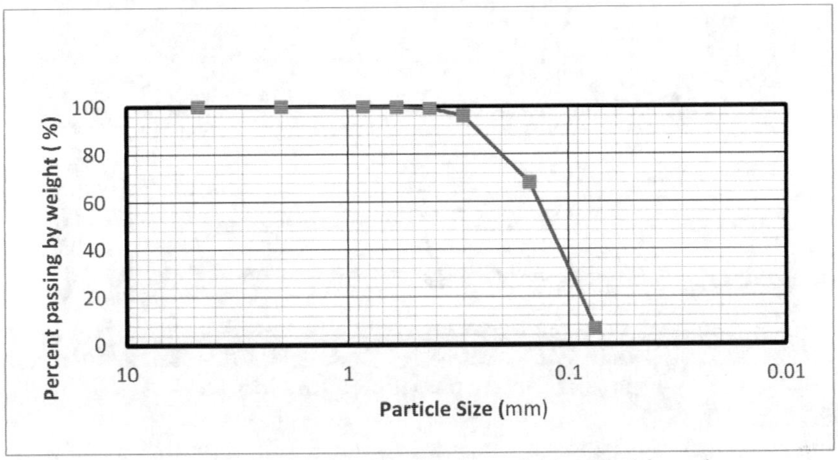

Figure 1. Grain size distribution of Abu Dhabi sand

Table 2. Index properties of the geofiber (as provided by the manufacturer)

Property	Test Method	Fiber Type
		Fibrillated
Material	ASTM C1116/C1116M	Synthetic macro fiber
Shape	-	Flat-Narrow (2mm wide)
Color	-	White
Moisture	-	None
Specific gravity	ASTM D792	0.92
Carbon black content	ASTM D1603	0.5%, minimum
Tensile strength (MPa)	ASTM D2256	620
Ignition point (C^{o})	-	590
Length used (mm)	Measured	40

2.2. Laboratory tests

California Bearing Ratio, CBR test was used to evaluate the strength and bearing capacity of the treated sand. Modified compaction test according to ASTM D1557 [12] was used to determine the maximum dry density (MDD) and optimum moisture content (OMC) of the sandy soil. The compaction curve of the modified Proctor test is shown in Figure 2. Sample preparation for CBR test is shown in Figure 3.

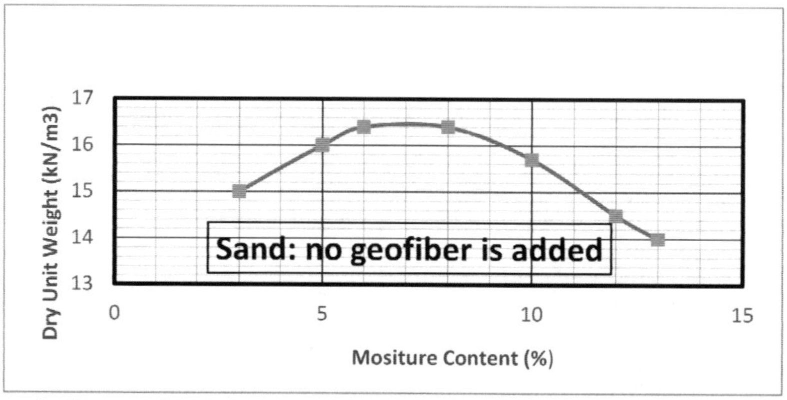

Figure 2. Compaction curve of the sandy soil according to modified Proctor test

Figure 3. Sample preparation for CBR test

The tests were performed on specimens prepared for CBR test according to ASTM D1883 [13] at OMC. The tested sand specimens included both; specimens without geofiber additives and specimens with varying geofiber content.

CBR test is commonly used in pavement design as an indicator of shear strength and bearing capacity of the soil. The CBR of a soil is the ratio obtained by dividing the stress required to cause a standard piston to penetrate 2.54 mm, 5.08 mm, 7.62 mm, 10.16 mm, and 12.70 mm into the soil by a standard penetration stress at each depth of penetration (ASTM D 1883). The CBR is an index value comparing the strength of the soil to that of crushed rock [14].

3. Results and discussion

The CBR tests were performed on samples contain three levels of geofiber; 0.1, 0.2, and 0.4 % by dry weight of the sand to evaluate the effect of the geofiber on the shear strength of the sandy soil. Also, CBR test was performed on sand specimen without addition of geofiber. Figure 4 shows that CBR values increased as the geofiber content increased and kept increasing within the amount of geofiber used in this study. CBR value increased from 19 to 55 at 0 and 0.4% geofiber respectively. This shows an improvement of 189% over the performance of unreinforced soil. Figure 5 shows comparison between the effect of varying geofiber content and different penetration depth. It shows that the CBR values has increased at all penetration depths as the geofiber content increased.

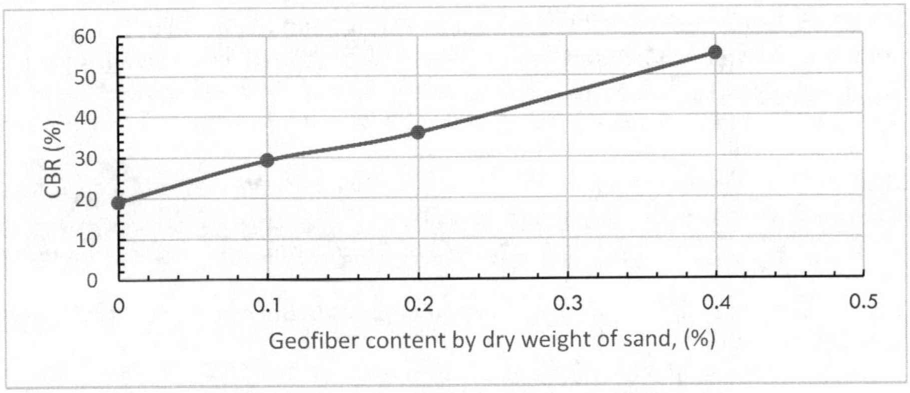

Figure 4. Effect of geofiber content on CBR values

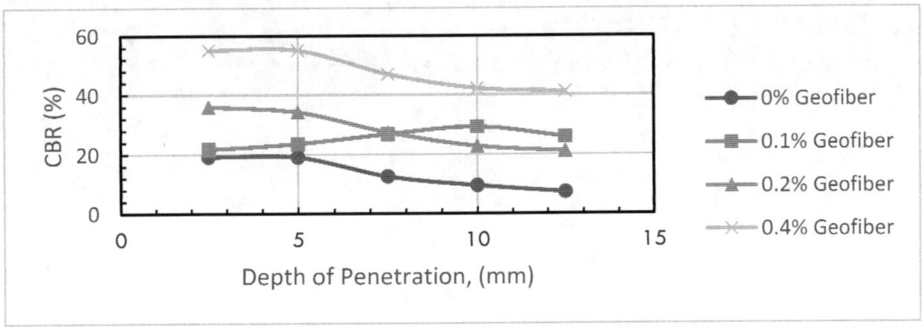

Figure 5. Comparative CBR performance of Abu Dhabi sand with Geofiber

4. Conclusion

Based on the results of this study, the shear strength of the treated soil has increased dramatically with addition of geofiber. No optimum value of geofiber content was observed. On the contrary, CBR values kept increasing in the range of geofiber content used in this study.

5. References

[1] Gray, D., and Ohashi, H. (1983), "Mechanics of Fiber Reinforcement in Sand", *Journal of Geotechnical Engineering*, Vol. 109, No. 3, pp. 335-353.

[2] Santoni, R. L., and Webster, S. L. (2001). "Airfields and road construction using fiber stabilization of sands." J. Transp. Eng., 127(2), 96e104.

[3] Liu, J., Wang, G., Kamai, T., Zhang, F., Yang, J., and Shi, B. (2011). "Static liquefaction behavior of saturated fiber-reinforced sand in undrained ring-shear tests." Geotext. Geomembr., 29(5), 462e471.

[4] Zhiwei Gao and Jidong Zhao, "Evaluation on Failure of Fiber-Reinforced Sand" Journal of Geotechnical and Geoenvironmental Engineering, Vol. 139, No. 1, January 1, 2013.

[5] Vasudev D. Performance studies on rigid pavement sections built on stabilized sulfate soils. Msc thesis, University of Texas at Arlington; 2007.

[6] Musenda C. Effects of fiber reinforcement on strength and volume change behavior of expansive soils. MS thesis, The University of Texas at Arlington, Arlington, Texas; 1999

[7] Puppala J, Musenda C. Effects of fiber reinforcement on strength and volume change behavior of expansive soils, trans res boa. In: 79th Annual meeting, Washington, USA; 2000

[8] ASTM C 128. *Standard test method for relative density (Specific Gravity) and absorption of fine aggregate.* West Conshohocken, PA.

[9] ASTM D 6913. *Standard test methods for particle-size distribution (Gradation) of soils using sieve analysis.* West Conshohocken, PA.

[10] Fletcher, C., & Humphries, K. 1991 California bearing ratio improvement of remolded soils by the addition of polypropylene fiber reinforcement. Transportation Research Record, 1295, 80–86

[11] Hazirbaba, K. 2019 Stabilization of aeolian sand with combined use of geofiber and synthetic fluid. Cogent engineering, 2019, 6

[12] ASTM D 1557. *Standard test methods for laboratory compaction characteristics of soil using modified effort.* West Conshohocken, PA.

[13] ASTM D 1883. *Standard test method for CBR (California Bearing Ratio) of Laboratory-Compacted Soils.* West Conshohocken, PA.

[14] Liu, C., & Evett, J. B. 2003 *Soil properties: Testing, measurement, and evaluation* (5th ed.). Prentice Hall: Pearson.

Acknowledgment

Author wishing to thank Abu Dhabi Department of Education and Knowledge (ADEK) for the financial support provided to this research.

IOP Conf. Series: Materials Science and Engineering **910** (2020) 012006 doi:10.1088/1757-899X/910/1/012006

Energy efficient, no firing, air binder. Recommendations for the processing of phosphogypsum waste

N F Bakhriev[1], Sh A Buriboev[1] and A Buriboev[2]

[1]Department of construction materials, Samarkand State Architectural and Civil Engineering institute (SSCI), Uzbekistan
[2]Department of Computer science, Konkuk University, Korea

Corresponding author's e-mail address: wolfsonabror@gmail.com

Abstract. The article presents data on the results of research and processing of phosphogypsum waste from the Samarkand Chemical Plant, over many years stored on a vast territory. Phosphogypsum contains up to 94% contains two water sulfates of calcium, which is the most valuable raw material to produce gypsum binders. However, phosphogypsum contains some chemical compounds that are harmful to human life and for this reason, it is difficult to use in industrial production without neutralization. The article presents the results of research and recommendations for solving this problem.

1. Introduction

Wall materials (bricks, stones, and blocks) account for about one-third of the total building cost of buildings and structures, including large Portland cement production, the most passionate of us all for the production or construction of durable bricks energy consumption. The main purpose of the chosen topic is to recommend and scientifically justify the use of energy-efficient materials based on modern, local raw materials, exploring the production of wall materials in terms of energy consumption.
As the solution to the main tasks of our scientific research the problem of production of gypsum-bonding agents and wall-based blocks is based on their solution.

Concentrated apatite concentrate contains phosphorous oxide P2O5 (in dry matter) up to 39%. Phosphorus extract is obtained when this raw material is treated with phosphorus and sulfate in some cases with nitric acid. In this process, the residue of phosphogypsum, which is many secondary products, is secreted. In recent years, 500 thousand tons of phosphogypsum waste more than a ton of waste is collected and waiting for recycling. The main ingredient in phosphogypsum is CaSO4 compound, which is 94% by weight [1,2]. If we analyze this material in accordance with the requirements of GOST 513-82, we can see that it is a type of gypsum. However, there are elements that prevent the use of phosphogypsum in the case of natural gypsum rock, which stops the use of this expensive product in the construction industry. It is usually a secondary product of the production of phosphogypsum, which is dissolved in powder, water-solubility does not exceed 0.2%, it is almost insoluble in water, and however, the harmful elements in the phosphogypsum are water-soluble, for example, dissolved in 0.5 to 1.5% water. Phosphorous oxide P2O5, 0.3-0.4% hydrogen phosphate ion (NPO4), and up to 1% of fluoride compounds, which must be neutralized before use. Washed in aqueous solution, it is believed that the harmful chemical elements in it are either neutralized or dissolved in lime water [3,4].

2. Proposed methodology

In order to fulfill the above objective and scientific hypothesis, the Samarkand Chemical Plant Superphosphate Fertilizer Plant was selected as a waste of phosphogypsum from the Soviet Union and occupy many areas. Even though this raw material has been stored for many years in the factory waste dumps, it is found that the elements dissolved in water-soluble, thermal treatment and milling of harmful gases - phosphorus hydrogen and phosphates.

IOP Conf. Series: Materials Science and Engineering **910** (2020) 012006 doi:10.1088/1757-899X/910/1/012006

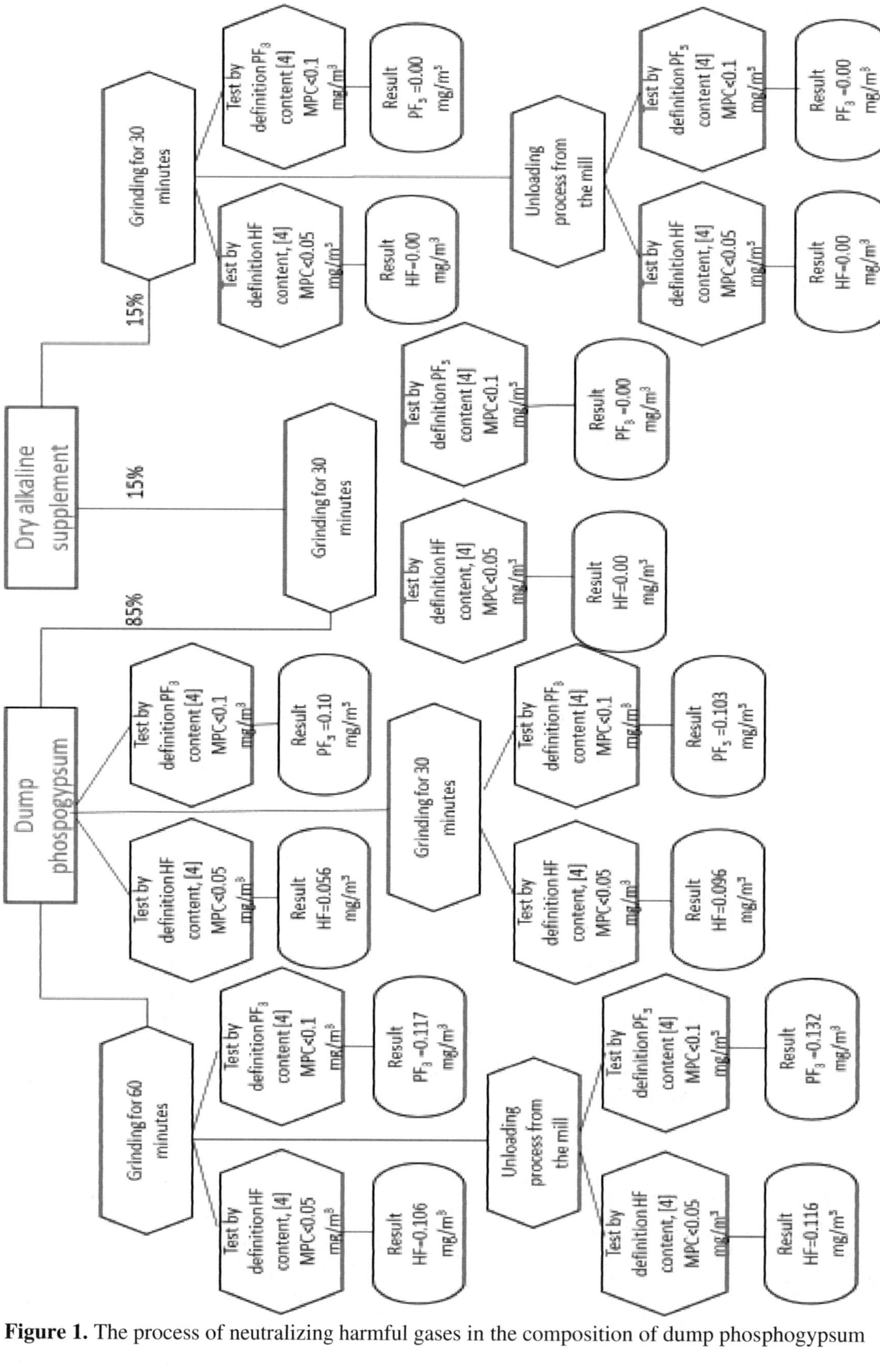

Figure 1. The process of neutralizing harmful gases in the composition of dump phosphogypsum

In order to implement the plans, set for our research, the task was to neutralize the harmful elements in the raw material during the milling process and to convert them into neutral compounds that are insoluble in water. In the process of scientific research and the analysis of emissions of the same type of industrial production, we have encountered alkaline dry waste of the Kungrad soda plant in Karakalpakstan. This industrial waste is a secondary product of calcium soda, sodium bicarbonate (food baking soda) and caustic soda, with 93% of active lime compounds. Many scientific works of literature cite the use of this secondary product as raw material, or as an additional element in the cement industry. However, we decided to study the physicochemical properties of this element in combination with phosphogypsum in our research.

The following were the objectives of the study:

- The problem of neutralization of fluoride hydrogen and phosphorus harmful gases in phosphate gases from the Samarkand Chemical Plant in milling with the dry alkaline residues of the Kungrad Soda Plant.
- For the purpose of this experiment, the Samarkand Chemical Plant was analyzed by emissions and extracts of phosphogypsum from various layers of long-term storage. In particular, a certain number of samples were extracted from the Kungrad Soda Plant's dry alkaline waste range. The chemical composition of these components was studied and presented in tables in the form of oxides.
- Methods of scientific research. The pre-planned internationally recognized method of neutralizing harmful gases in phosphogypsum has been adopted. Detection and detection of hazardous amounts of fluoride hydrogen gas are followed by a 15% solution of calcium and glycerin containing 3% carbon dioxide as figure 1, [5,6]

3. Experiments, results

The experiments were conducted in a closed building in the department of grinding and mills of Regional Research Laboratory-5 of SSCI. The aim was to observe the changes in the laboratory microclimate and samples when the phosphogypsum was pulled in a mill when the harmful fluoride hydrogen or phosphorus hydrogen gas was removed, and the phosphorous substance was mixed in a mill with a dry alkaline component.

The method of determining the amount of hydrogen phosphorus in the air or in the body is based on the determination of the concentration of gas by phosphoric acid as a result of oxidation of potassium permanence solution. Studies were performed at the lab site, during the drying of the items, in the milling process, and at half-time intervals of samples (table 1).

The use of sections to divide the text of the paper is optional and left as a decision for the author. Where the author wishes to divide the paper into sections the formatting shown in table 1 should be used.

Table 1. Formatting sections, subsections and subsubsections.

Components, %		Place of analysis and sampling period	Phosphorus-hydrogen (PH_3) - the result of analysis from two parallel layers, mg / m^3	Fluoride Hydrogen (HF) - the result of analysis from two parallel layers, mg / m^3
Dump phosphogypsum	Dry alkaline supplement			
First layer test, 100	0	At the time of loading in the mill	0.01	0.056
		30 min after grinding in a mill	0.09	0.09
		60 min after grinding in a mill	0.093	0.097
		At the moment of emptying the mill	0.28	0.135

IOP Conf. Series: Materials Science and Engineering **910** (2020) 012006 doi:10.1088/1757-899X/910/1/012006

Second layer test, 100	0	At the time of loading in the mill	0.01	0.056
		30 min after grinding in a mill	0.103	0.094
		60 min after grinding in a mill	0.098	0.088
		At the moment of emptying the mill	0.13	0.099
Third layer test, 100	0	At the time of loading in the mill	0.10	0.056
		30 min after grinding in a mill	0.103	0.096
		60 min after grinding in a mill	0.118	0.106
		At the moment of emptying the mill	0.132	0.117
Third layer test, 95	5	At the time of loading in the mill	0.10	0.056
		30 min after grinding in a mill	0.05	0.03
		60 min after grinding in a mill	0.017	0.02
		At the moment of emptying the mill	0.009	0.01
Third layer test, 90	10	At the time of loading in the mill	0.01	0.056
		30 min after grinding in a mill	0.003	0.001
		60 min after grinding in a mill	0.001	0.00
		At the moment of emptying the mill	0.00	0.00
Third layer test, 95	15	At the time of loading in the mill	0.01	0.056
		30 min after grinding in a mill	0.00	0.00
		60 min after grinding in a mill	0.00	0.00
		At the moment of emptying the mill	0.00	0.00

Note:

- fluoride hydrogen (HF) - maximum permissible concentration of atmospheric air (PDC)- 0.05 mg / m^3;
- Phosphorus hydrogen (PH_3) - maximum permissible concentration of gas in atmospheric air (PDC) - 0.1 mg / m^3.

4. Discussions

According to our experimental prohibitions, a small amount of fluoride hydrogen and phosphorus hydrogen gases were detected in the samples from various layers of phosphogypsum, which have been stored for many years at the Samarkand Chemical Plant. In the deep layers of storage, there is practically no change in content, the maximum permissible concentration of harmful gases is more than 2 times. The main studies were carried out in 3 layers (depth up to 2 meters). It was observed that during the drying process, with the addition of 5% dry alkaline additives to the phosphogypsum, the number of harmful gases had a permissible concentration limit, and the gases were neutralized when the mixture was milled. It was found that no increase in harmful gases was observed in lab air and mill dust when the number of dry alkaline additives increased by 15%. Based on the repeated experiments, it is recommended to use the dry alkaline additives at the Kungrad Soda Plant with 10-15% in the process of thermal treatment of phosphogypsum and milling.

5. References

[1] Greene M 2009 *Gypsum Construction Handbook*, 6 ed. Construction Industry by CGC Inc., RS Means Company, Kingston, MA, 556 pp.

[2] Sampson D H 2011 *Gypsum: Properties, Production and Applications*. Nova Science Publishers Inc., New York, 371 pp.

[3] Lutskiy D Litvinov T Ignatovich A Fialkovskiy A 2018 *Complex Processing of Phosphogypsum* – A way of recycling dumps with reception of commodity production of wide application. Journal of Ecological Engineering. 19(2), 221–225.

[4] Lushnikova N and Dvorkin L 2016 *The sustainability of lightweight aggregates manufactured from clay wastes for reducing the carbon footprint of structural and foundation concrete. In book: Sustainability of Construction Materials.* Ed. by. J. Khatib. Cambridge: Woodhead Publishing, 643-681 pp.

[5] RD 52.04.797-2014. Mass concentration of hydrogen fluoride in samples of atmospheric air. Measurement procedure by the photometric method using xylenol orange.

[6] Guidelines for photometric determination of hydrogen phosphorus in air. Ministry of Health of the USSR, M: CRIA, Morflot, 1981

Acknowledgments

Authors wishing to acknowledge assistance or encouragement from colleagues, special work by technical staff or financial support from organizations should do so in an unnumbered Acknowledgments section immediately following the last numbered section of the paper.

Experimental investigation into the failure process of exterior beam-column joints with high-strength reinforcements

H Arshadi[1], A Kheyroddin[2], H Naderpour[3], and *M Kioumarsi[4]

[1, 2, and 3] Faculty of Civil Engineering, Semnan University, Semnan, Iran.
[4] Faculty of Technology, Art and Design, Department of Building and Energy Technologies, Oslo Metropolitan University, Oslo, Norway

*mahdik@oslomet.no

Abstract. Although using high-strength reinforcements (HSRs) have economical and execution benefits, their possible negative effects on the performance of structures such as ductility, energy dissipation, and damage process caused them to be restricted by building codes in seismic areas. In this research, the HSR effects on the cyclic behavior, cracking process, and damage indices of exterior beam-column joints were experimentally investigated under cyclic loading. Three exterior beam-column joints with HSRs (with the yield strengths of 500 and 580 MPa) were designed based on the special seismic provisions of ACI 318-19. These specimens were subjected to cyclic loading and their responses were obtained. The results showed that using HSRs led to increasing the width and depth of the cracks. Then, two methods of calculating damage indices (introduced by Promis *et al.*, 2009) were applied based on the experimental results. Considering the damage indices, it was observed that using HSRs increased the damage indices of the specimens and the damage distribution in them.

1. Introduction

Several structural systems resist earthquake-induced forces such as shear walls, moment frames, and dual systems (moment frames with shear walls). Although there are many studies on performance of the lateral resisting systems [1-5], but these systems have some usage limitations such as height of structures, etc. [6]. For instance, special moment frames along with shear walls are earthquake-resisting systems that high-strength reinforcement (HSR)–steel reinforcements with the yield strengths greater than 500 MPa–application is restricted in them due to possible repercussions of HSR application [7,8]. There are considerable challenges against the HSR application such as increasing the crack width, the brittle failure, decreasing the ductility and energy dissipation. On the other hand, HSR application has several benefits, such as diminishing the labor and material costs, reducing construction time and facilitating the construction procedure [6].

Damage indices can be calculated based on different parameters such as strain, stress, displacement, dissipated energies, stiffness, and ductility. Several researchers have proposed Damage indices (DI) for different types of structures during recent years. These indices have been modified and calibrated by passing the time. Promis *et al.* (2009) carried out seismic damage analysis for reinforced-concrete (RC) buildings

[9]. They proposed and used several damage indices in their study. The damage indices can be studied by analytical methods like incremental dynamic analysis (IDA) and endurance time (ET) method. In these methods, structures are subjected to incremental cyclic loads and their responses are calculated until the damage indices get certain amounts. Arshadi (2016) went over the concepts and methodologies of the IDA with single and multiple records [10]. Also, Arshadi *et al.* (2014) compared these two methods in the frames rehabilitated by base isolators [11]. The initial experiments on the RC members with HSRs did not include cyclic loading. Firstly, cyclical experiments were carried out by Burns and Siess to study the performance of the elements subjected to seismic excitations in the early 1960s [12]. Kheyroddin *et al.* (2017) studied the HSR effects on the cracking, drifts, decreasing steel consumption and effective moment of inertia of beams and columns by nonlinear static and dynamic analyses in the intermediate moment frames [12]. Their research showed that HSR application increased the concrete cracks and decreased the effective stiffness of RC structures. Arshadi *et al.* (2019) experimentally studied the behavior of frames and beam-column joints with HSRs subjected to cyclic loading [6, 13]. They focused on different parameters such as energy absorption, stiffness degradation, cracking and plastic hinge formation in their research. They showed that using HSRs led to a considerable decrease in ductility and energy absorption of the specimens. Dabiri and Kheyroddin (2020) investigated the effects of bars on the behavior of non-seismically detailed RC beam-column joints [14]. Arshadi *et al.* (2020) experimentally investigated the damage indices of special moment frames with HSR bars. They showed that HSRs expedited the failure process of the frames [15].

In this study, the HSR influences on the failure process and damage indices of three exterior beam-column joints were studied. Three beam-column joints were designed based on the special seismic provisions of ACI318-19 [16]. The specimens were constructed with the same geometry and equivalent amounts of the reinforcements. The specimens were tested under cyclic loading. The experimental results showed that the HSR application decreased the cracking forces and increased the crack propagation in the specimensBesides, the damage indices introduced by Promis *et al.* (2009) were used to study the damage behavior of the joints [10]. One of these methods was based on the displacements and the other one was based on the energy absorption. The slope of damage diagrams indicates the velocity of damage distribution in the specimens. The damage indices showed that the velocity of the damage propagation in case of applying HSRs as both the longitudinal and transverse reinforcements was the most among the beam-column joints. Also, applying HSRs as longitudinal reinforcements decreased the velocity of damage propagation in comparison to applying them as just the stirrups.

2. Experimental program

2.1. Material properties

Normal ready-mixed concrete with the water-cement ratio of 0.43 and Portland cement type 2 was used to achieve a compressive strength of 30 MPa and to cast the specimens. As for HSRs, two types of HSRs with the yield strengths of 500 and 580 MPa produced by the Thermex method were used. The Thermex method is just a swift water quenching process that ensures consistent properties over the entire bar length [6].

2.2. Specifications of the specimens

An exterior beam-column joint is a part of a moment frame and achieved by cutting through points of contra-flexure of the beams on both sides of the column and cutting through the column nearly one-half story height above and below the joint. Studying the behavior of beam-column joints with HSRs is recommended by the technical reports [13]. In this paper, three exterior beam-column joints (nearly the two-thirds scale of the real one) with the same geometry and concrete properties were designed based on the special seismic instructions of ACI 318-19 [16]. The specifications of the columns were identical in all the specimens because it was intended that the columns had to remain elastic. These columns which played the role of the foundation for the beams had the same steel bar arrangements and were built with the 500 MPa steel. In the specimens, HSRs were only used in the beams. Firstly, the reference model whose both transverse and longitudinal reinforcements were the 500 MPa steel, was designed. Then, the other two beam-column joints with the equivalent 580 MPa steel as the longitudinal or transverse reinforcements were designed based on the reference specimen [6]. Table 1 shows the names and properties of the test models.

Table 1. Properties of the longitudinal and transverse steel bars in the beam-column joints

Models	Yield stress of longitudinal rebars, f_{yl} (MPa)	Yield stress of transverse rebars, f_{yt} (MPa)
BL500S500	500	500
BL580S500	580	500
BL580S580	580	580

2.3. Test procedure

The lateral cyclic load was applied to the top of the beam by a bidirectional hydraulic jack with the loading capacities of 2000 kN in compression and 1000 kN in tension [6]. The specimens were fixed to the rigid floor by the metal jackets. This actuator was attached to the beams by a system consisted of two plates joined to each other by two ultra-high-strength rods. All the specimens were tested under cyclic lateral loading at the top of the beams. Figure 1 shows the laboratory setup of the specimens. The protocol employed in horizontal cyclic loading was a displacement control one proposed by ACI 374 [13].

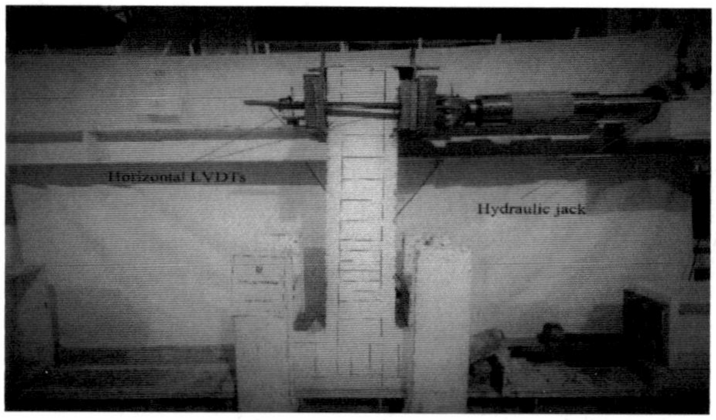

Figure 1. Laboratory setup of the beam-column joints [8]

3. Experimental results

3.1. Crack patterns

The initial cracks were moment ones formed near the end of the beam (close to their joint to the columns). In the BL500S500 and BL580S580 specimens, the initial cracks were formed at the 0.12% drift level and in the BL580S500 specimen, the first cracks were formed at the 0.14% drift level. The number, width, and depth of the shear and moment cracks increased by applying more loading [6]. However, no cracks were formed on the columns, because of the weak beam-strong column rule considered in the designing. The average cracking loads (in tension and compression) for the BL500S500, BL580S500 and BL580S580 specimens are 6, 5.5 and 3.75 kN, respectively. As for the crack distribution, the higher the yield strength of the steel, the wider and deeper the cracks.

3.2. Indices based on displacements

The displacement-based damage indices used in this research were proposed by Promis *et al.* (2009) [9]. This index which indicates the linear accumulation of the damage is calculated as follows:

Firstly, the cumulative term β_ω is calculated:

$$\beta_\omega = C\sum_i \frac{\delta i}{\delta f} \qquad (1)$$

where δ_i is the maximum displacement for cycle i, δ_y is the displacement corresponding to the yield of steels, δ_f is the failure displacement in monotonous loading, and C = 0.1. Then the total damage D is calculated based on β_ω. The values of this damage index function in extreme cases are: f (0) = 0 and f (1) = 1. The definition of DI (damage index) according to β_ω is:

$$DI = \frac{e^{n\beta\omega} - 1}{e^{n} - 1} \qquad (2)$$

In which: n is considered 1 for the strongly reinforced nodes, otherwise, n = -1 [9]. The displacement damage indices of the specimens calculated by the above definition are shown in Figure 2. It must be mentioned that the slope of damage diagrams shows the velocity of damage distribution. In the beam-column joints, the damage propagation velocity of the BL500S500 specimen (based on its damage diagram slope) was more than the other two specimens. Moreover, the BL580S580 and BL580S500 specimens had similar displacement damage propagation. The BL500S500 specimen also had the highest damage capacity among the beam-column connections. This means that it endured greater amounts of forces and failed later than the other specimens because of less intensity of damage distribution. The BL580S500 had the least damage capacity and failed sooner than the other specimens.

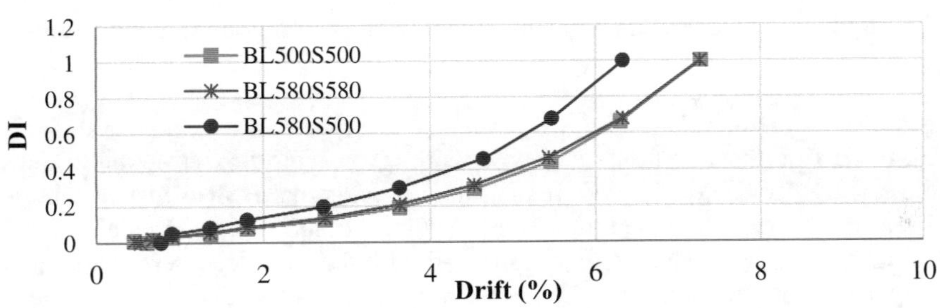

Figure 2. Displacement-based damage indices of the beam-column connections

3.3. Indices based on energy

Promis *et al.* (2009) proposed a formulation for damage index based on the dissipated energy of the RC structures [9]. This formulation depends on the dissipated energy, but just during cycles for which the force of displacement is higher than that of the one related to the steel yielding. This means that just the cycles in which $\frac{Fi}{Fy} \geq 0.75$ are taken into account. In which F_i is the force reached during the cycle i, and F_y is the yield force of the RC structures. The formulation to calculate the damage index is as below:

$$DI = \sum_i \frac{Fi\delta i}{Fy\delta y} \qquad (3)$$

where δ_i & F_i are the maximum displacements and its corresponding forces for cycle i, δ_y & F_y are the displacements related to the steel yielding and its corresponding forces. This definition of damage index depends on the determination of specimen failure, which happens at a 25% decrease in the maximum load of the hysteresis loops. It is worth mentioning that the relation to calculate the dissipated energy, according to the surfaces of the hysteresis loops is as follows:

$$\int dE=\int F(\delta)d\delta = \int M(\phi)d\phi \qquad (4)$$

where E is the dissipated energy, M (ϕ) & ϕ are the moment and its corresponding rotation, F (δ) & δ are the force and its corresponding displacement. Figure 3 indicates the damage indices in the beam-column joints. In the beam-column joints, the energetic damage index of the BL580S500 specimen was less than the other specimens. As for this damage index definition, the previous legacy which was shown in the displacement-based damage indices was observed, too. This means that the BL500S500 had the best performance in terms of damage propagation and the BL580S500 had the least damage capacity among the other specimens.

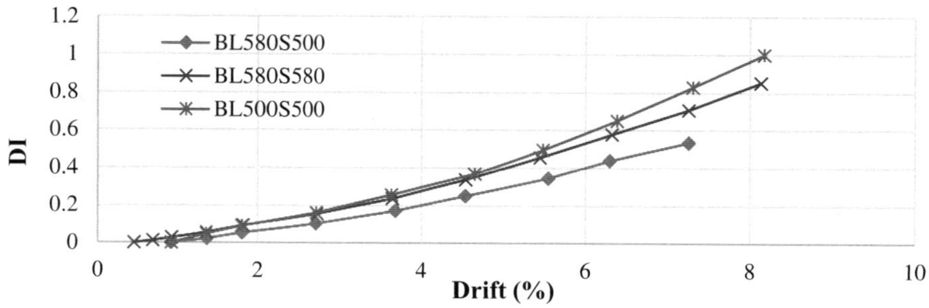

Figure 3. The energetic damage indices of the beam-column connections

4. Conclusion

In this research, the HSR effects on the failure process of three exterior beam-column joints were studied. The beam-column joints are one of the important structural parts that HSR effects on them are not studied sufficiently. Then, three beam-column joints were designed based on the special seismic provisions of ACI318-19 in this research [16]. They were constructed with the same geometry and tantamount amounts of the reinforcements. The specimens were subjected to cyclic loading. The experimental results indicated that the HSR application decreased the cracking forces and increased the crack propagation in the specimens. Then, the damage indices introduced by Promis *et al.* (2009) [9] were applied to investigate the damage behavior of specimens. One of these methods was based on the displacements and the other one was based on the energy absorption. The damage indices showed that the velocity of the damage distribution in the case of applying HSRs as just the longitudinal reinforcements (BL580S500) was the most among the specimens. It must be reminded that the slopes of the damage diagrams show the velocity of damage distribution in the specimens. Moreover, applying HSRs as both the longitudinal and transverse reinforcements (BL580S580) decreased the velocity of damage distribution in comparison to applying them as just the longitudinal ones (BL580S500). This means that higher-strength steel reinforcements as longitudinal and transcerse ones had better interaction with each other than using high-strength and low-strength ones with each other. Finally, the BL500S500 specimen had the best damage performance and failed in a higher drift ratio than the other specimens.

5. References

[1] Kioumarsi, B., Gholhaki, M., Kheyroddin, A. and Kioumarsi, M., 2016. Analytical study of building height effects over Steel Plate Shear Wall Behavior. International Journal of Engineering and Technology Innovation, 6(4), p.255.

[2] Bypour, M., Gholhaki, M., Kioumarsi, M. and Kioumarsi, B., 2019. Nonlinear analysis to investigate effect of connection type on behavior of steel plate shear wall in RC frame. Engineering Structures, 179, pp.611-624.

[3] Moradi, M. J., Roshani, M. M., Shabani, A., & Kioumarsi, M. (2020). Prediction of the Load-Bearing Behavior of SPSW with Rectangular Opening by RBF Network. *Applied Sciences*, *10*(3), 1185.

[4] Kheyroddin, A., Kioumarsi, M., Kioumarsi, B., & Faraei, A. (2018). Effect of lateral structural systems of adjacent buildings on pounding force. *Earthquakes and Structures*, *14*(3), 229-239.

[5] Bypour, M., Kioumarsi, B., & Kioumarsi, M. (2019). Investigation of Failure Mechanism of Thin Steel Plate Shear Wall in RC Frame. In *Key Engineering Materials* (Vol. 803, pp. 314-321). Trans Tech Publications Ltd.

[6] Arshadi H, Kheyroddin A, and Naderpour H, 2019, High-strength reinforcement effects on the seismic behaviour of beam-column joints, Proceeding Institute of Civil Engineering - *Structures and Buildings*, doi:10.1680/jstbu.18.00225.

[7] Arshadi H and Kheyroddin A, 2019, Shear lag phenomenon in the tubular systems with outriggers and belt trusses', *Magazine of Civil Engineering*, **86**(2): 105–118, doi: 10.18720/MCE.86.10.

[8] Kheyroddin A, Mirrashid M and Arshadi H, 2017, An investigation on the behavior of concrete cores in suspended tall buildings, *Iranian Journal of Science and Technology - Transactions of Civil Engineering*. Springer International Publishing, 41(4), pp. 383–388. doi: 10.1007/s40996-017-0075-y.

[9] Promis G, Ferrier E, and Hamelin P, 2009, Effect of external FRP retrofitting on reinforced concrete short columns for seismic strengthening, *Composite Structures*, **88**:367–79. doi:10.1016/j.compstruct.2008.04.019.

[10] Arshadi, H, 2016, An overview on the concepts and methodologies of incremental dynamic analysis IDA (with a single record and multiple records). The 2016 Structures Congress (ASEM), Jeju Island, Korea.

[11] Arshadi H, Izadifard, R A, and Estekanchi H, 2014, The comparison of the IDA and ET on steel structures rehabilitated by isolators, *Journal of Rehabilitation in Civil Engineering*, **2 (2)**: 46-62. doi: 10.22075/jrce.2014.208.

[12] Kheyroddin A, Arshadi H, and Binaipur F, 2017, An overview of the effects of high-strength reinforcement (HSR) on the intermediate moment-resisting frames, *AUT Journal of Civil Engineering*. doi:10.22060/ceej.2017.12406.5207.

[13] Arshadi H, Kheyroddin A, and Naderpour H, 2019, An investigation into the behavior of special Moment frames with high-strength reinforcement Subjected to cyclic loading. *Journal of Building Engineering,* **26**:100905. doi:10.1016/j.jobe.2019.100905.

[14] Dabiri H. and Kheyroddin A, 2020, Influence of reinforcement on the performance of non seismically detailed RC beam-column joints. *Journal of Building Engineering*, doi: 10.1016/j.jobe.2020.101333.

[15] Arshadi, H, Kheyroddin, A, Naderpour, H, Kioumarsi M, 2020, Experimental study on the failure traits of frames with high-strength reinforcements', *International Journal of Materials Science and Engineering (IJMSE)*, (in press).

[16] American Concrete Institute (ACI), 2019, ACI 318-19: Building code requirements for structural concrete, American Concrete Institute, Farmington Hills, MI, USA.

IOP Conf. Series: Materials Science and Engineering **910** (2020) 012008 doi:10.1088/1757-899X/910/1/012008

Comparative analysis of the seismic response of an essential and a common building using site spectra and dynamic modal spectral analysis according to South American standards

F Huaripata[1], J F Torres[1] and J Orihuela[1]

[1] School of Civil Engineering, Peruvian University of Applied Sciences, Peru

Corresponding author's e-mail: u201517140@upc.edu.pe

Abstract. The buildings built on coastal soil, located on the fire belt present a high seismic risk, since there are unfavorable soil conditions and in other cases plate subduction. This has led to countries located in these areas developing seismic regulations that allow the proper design of engineering works; however, various regulations present different considerations for the design, such as the use of parameters that allow standardizing the design that often does not present the same seismic responses of the real dynamic behavior of the soil. Therefore, the present research proposes to evaluate the seismic response of two buildings, one essential and the other common, with an analysis using site spectra, a dynamic spectral modal analysis and additionally a time-history analysis taking into account soil types and use of different seismic parameters of South American standards. The main result is that the seismic response by analysis with site spectra (ES) is greater compared to the dynamic spectral modal analysis according to the magnitude of the earthquakes and the type of soil.

1. Introduction

The western coast of South America presents a constant and high seismic activity, this due to the subduction of the Nazca plate below the South American plate. In this way, the affected countries have strong institutions that analyze the tectonic behavior of these plates. The study carried out by Tavera [1] shows that in the central part of the South American west coast there are 5 zones with a high concentration of seismic energy. It is important to emphasize that in several countries of South America there has been a seismic silence for several years that increases their vulnerability.

Several authors suggest that the design spectrum for a building should be done taking into account the site parameters. Eskandarinejad et al. [2] propose that for the use of design spectra the propagation of seismic waves through the soil layers should be considered, since the spectral content and the characteristics based on the amplitude of the movements of the terrain depend on them. On the other hand, Lagaros et al. [3] consider that a series of seismic registers correctly selected for obtaining seismic capacity of the structural system through linear or non-linear seismic analysis should be taken into account. In addition, Papaspiliou et al. [4], propose to consider soil amplification and probabilistic seismic risk calculations (PSHA) for seismic analysis with site spectra.

Finally, Sonmezer et al. [5] developed a comparative study of seismic performance through a response history analysis of 3 existing structures according to European regulations and site spectra (ES). The results obtained show that the use of the site spectrum, compared to the code spectra, had a great effect on the performance of the irregular structure.

The present investigation proposes to make a comparison between an analysis with site spectra and a dynamic modal analysis in two buildings of different category of importance, mainly considering South American standards. One of the structures will be hospitable and the other a multifamily housing. The main objective of this comparative analysis is to evaluate the effect generated by site spectra on the structural seismic response.

2. Method

In the investigation that is going to be raised, the comparison of seismic responses of reinforced concrete buildings in different conditions will be carried out through seismic analysis methods such as spectral modal analysis, taking into account the importance of the building following the proposed guidelines by Peruvian seismic regulations E.030 and Chilean regulations NCh433; In addition, the other method to be used will be to perform a seismic analysis with site spectra, for which records of the earthquakes of Maule 2010 E-O, Arequipa 2001 N-S and Lima 1974 E-O will be used.

2.1. Structural modeling

The structural comparison of the structures was carried out: an essential and a common one. For this, the use of computer software for structural analysis was used and the loading conditions of the South American regulations were taken into account, however, other considerations can be taken into account, depending on where the building is carried out. The modeling was done in plan and a 3D view that helps to have a better visualization of the structures.

2.2. Design spectra according to regulations

The structural comparison of the structures was carried out: an essential and a common one. For this, the use of computer software for structural analysis was used and the loading conditions of the South American regulations were taken into account, however, other considerations can be taken into account, depending on where the building is carried out. The modeling was done in plan and a 3D view that helps to have a better visualization of the structures.

2.2.1. Peruvian standard (E.030). The design spectrum of the structures, seismic parameters were used, which were obtained based on the seismic-resistant E.030 standard. The design spectrum of Peruvian regulations is represented by the pseudo-acceleration and the period of the structure. In the investigation, 8 design spectra were used in total, since there are two degrees of importance for the type of building, two axes of analysis and two soil conditions, however only the most significant ones that are related to S2 soil are evaluated.

2.2.2. Chilean standard (NCH433). The design spectrum obtained under the conditions of NCH433, is quite similar to the previous regulations, however the graph is different because of the difference that for the soil amplification parameter ($\propto$) and the reduction factor R * is They use formulas that depend on the fundamental period of vibration of the structure and the soil conditions. The number of spectra used was the same as the previous one and in the same way only the most representative ones will be evaluated.

2.3. Site spectrum

For the development of one of the seismic analyzes, it is necessary to obtain site spectra that consider the true behavior of the soil. Alva et al. [6] developed a geotechnical study to obtain the dynamic response of the soil on the surface considering the behavior of each type of soil. For this they used the earthquakes of Maule 2010 EO, Arequipa 2001 NS and Lima 1974 EO, scaled them to a uniform hazard spectrum of 500 years and based on cutting wave speeds, degree of soil damping this record was scaled from the rock to the surface, resulting in the Site Spectrum. The Maule earthquake had a magnitude of 8.8 degrees on the Richter scale with an epicenter in the Pacific Ocean and shook the southern center of Chile, for this earthquake the EO direction was taken, the Arequipa earthquake had a magnitude mb equal to 6.9, mainly affected the city of Arequipa, for the consideration of this earthquake the NS direction was taken. As for the Lima earthquake of 1974, with intensity of 8 degrees on the Mercalli scale and magnitude 5.6 on the Richter scale, the epicenter went to southern Lima and affected the districts of Chorrillos, La Molina, Cercado de Lima and the Constitutional Province of Callao, the registration at the EO address was used.

In the investigation a modal analysis was considered considering these spectra, which are in Figure 1, comparing it with spectra of the E.030 standard of the essential building, to determine the graphic difference between these spectra.

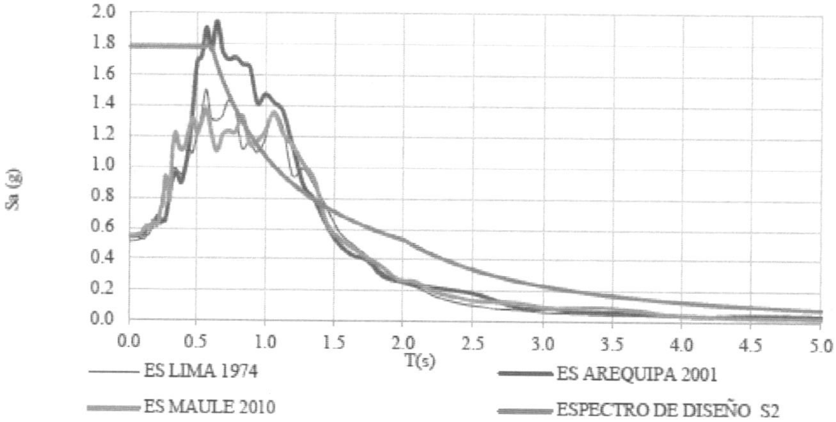

Figure 1. Site spectrums used in an elastic range

2.4. Seismic analysis

The modal spectral analysis was carried out following the guidelines of the South American regulations and following the methodology of Chopra [7]. For this, computational design and structural analysis software and the mentioned design spectra were used, as can be seen in Figure 1 with the design spectrum S2.

The Analysis with site spectra was performed following the modal analysis methodology, but considering other spectra, called site spectra (ES). For this analysis, the ES plotted in Figure 1 were used and computational software was used to obtain the seismic response of the buildings. In the present investigation the analysis of structures whose period is in 0.5 and 1 seconds has been prioritized, since it is where the ES presents a greater spectral acceleration.

3. Method

3.1. Existing structures

In order to make a comparison of structural results between ES and South American standards, a seismic analysis is carried out in a hospital building and a multi-family building with the help of the Etabs software. The seismic parameters of standard E.030 and Nch433 are taken. The structural representation of the essential and common building are shown in Figures 2 and 3, respectively.

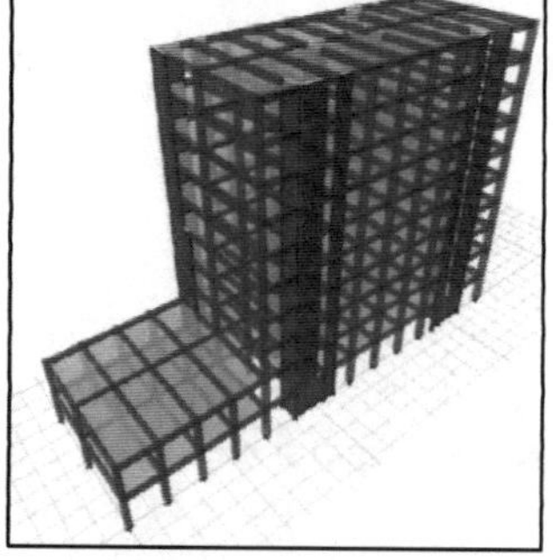

Figure 2. Hospital building

IOP Conf. Series: Materials Science and Engineering **910** (2020) 012008 doi:10.1088/1757-899X/910/1/012008

Figure 3. Common building

The fundamental vibration periods of the hospital and multifamily structure are 0.71 s and 0.52 s for the X direction, respectively, 0.98s and 0.32 s in the Y direction. As for the seismic parameters set forth in E.030 and Nch433, they will be considered the following values.

3.2. Design spectrum

To carry out the design spectra, the parameters present in Tables 1 and 2 were taken into account considering a type of soil S2 and D for both Peruvian and Chilean regulations. On the one hand, the design spectra for the 2 types of buildings, according to Peruvian regulations, are presented in Figure 4, where the initial values of spectral acceleration are between 0.20g and 0.45g. On the other hand, in Figure 5, the design spectra according to the Chilean standard are shown, which are different in shape due to the different calculation method, the initial spectral acceleration values are between 0.10g and 0.35g.

Table 1. Parameters according to standard E.030 for the seismic analysis

Parameters	Characteristic	Common building		Essential building	
Zone Factor	Zone 4	Z=	0.45	Z=	0.45
Category and use factor	Category C and A	U=	1.0	U=	1.5
	Soil, Type S2	S=	1.05	S=	1.05
Site Parameters	Periods	T_p=	0.60	T_p=	0.60
		T_L=	2.00	T_L=	2.00
Basic coefficient of reduction of seismic forces	Structural system Reinforced concrete: dual	R_x=	5.4	R_x=	6.3
		R_y=	5.4	R_y=	4.2

Table 2. Parameters according to standard NCH433 for the seismic analysis

Parameters	Characteristic	Common building		Essential building	
Zone Factor	A_0	A_0=	0.40	A_0=	0.40
Importance factor	Category II and IV	U=	1.0	U=	1.5
Soil factor	Soil type D	S=	1.20	S=	1.20
		T_o=	0.75	T_o=	0.75
		T'=	0.85	T'=	0.85
		n=	1.80	n=	1.80
		p=	1.00	p=	1.00
Response Modification Factor	Reinforced concrete	R_o=	11	R_o=	11
Dynamic Reduction Factor	R*	R^*_x=	4.10	R^*_x=	4.80
		R^*_y=	3.14	R^*_y=	5.69

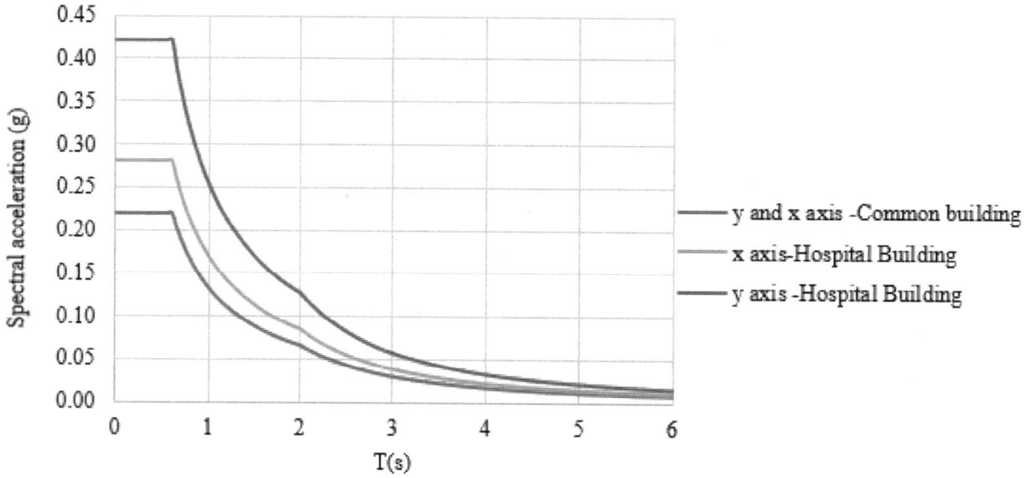

Figure 4. Design spectrums for buildings according to E.030

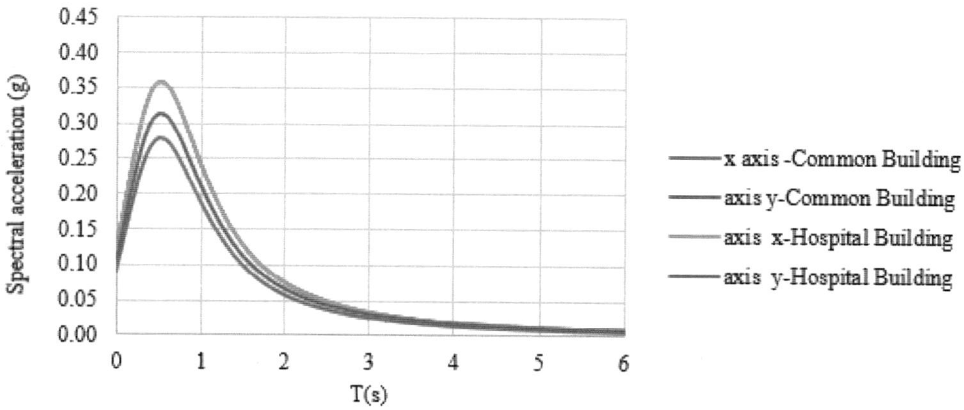

Figure 5. Design spectrum for buildings according to NCH433

3.3. Drifts and comparison

In the first place, static analysis is carried out, in order to identify the various irregularities existing in both structures, identifying for the hospital the Irregularity of Torsional and Mass; and for the multi-family corner incoming.

To conclude with the seismic analysis, we proceed to perform the dynamic spectral modal analysis according to the guidelines of the standards, with which the values of the mezzanine drifts of the building are obtained.

It can be seen in Figure 6 that the drifts of the multifamily building comply with the limit of 0.007 proposed by the E.030, obtaining greater displacements in the ES of Arequipa 2001. However, the result in Figure 7 is all contrary to the hospital building, since in the drifts of the Peruvian norm it is above the ES and the Chilean norm. This demonstrates that the Use factor present in both standards has a great impact on the results.

In addition to the relative displacements, the drift relations of the buildings were also obtained to represent the effects of the local spectrum in terms of comparison with those obtained by the South American seismic norms. As shown in Table 4 for multifamily building, it was determined that the maximum drift of the ES represents 145% and 142% of the E030 and NCH433, respectively. However, according to Table 3, in hospital construction, the ES compared to the Peruvian norm represent 90% of its drifts, otherwise the Chilean norm shows where the ES represent 148% of the respective standard.

IOP Conf. Series: Materials Science and Engineering **910** (2020) 012008 doi:10.1088/1757-899X/910/1/012008

Table 3. Comparative analysis of ES / E.030 drifts in the hospital building

Floor	ES Arequipa 2001	ES Lima 1974	ES Maule 2010	Average
9	0.91	0.73	0.76	0.80
8	0.92	0.73	0.77	0.80
7	0.93	0.74	0.78	0.81
6	0.94	0.75	0.78	0.82
5	0.95	0.75	0.79	0.83
4	0.95	0.76	0.79	0.83
3	0.95	0.76	0.79	0.83
2	0.95	0.75	0.79	0.83
1	0.94	0.75	0.79	0.82
0	0.93	0.74	0.78	0.82

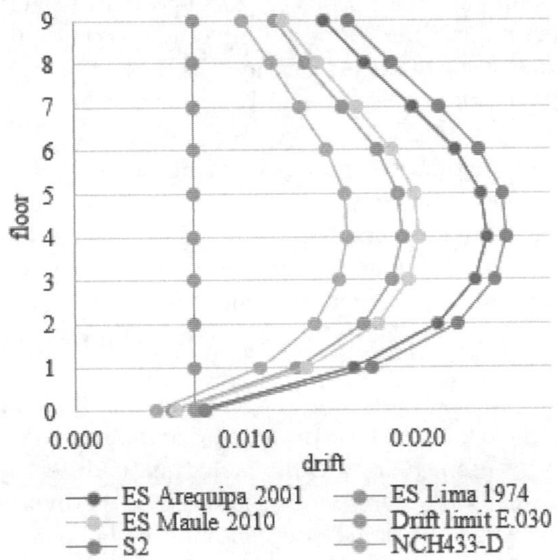

Figure 6. Drift of hospital building on the y axis

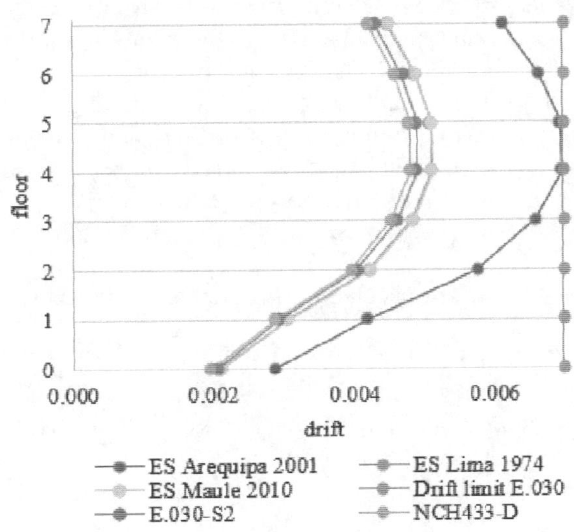

Figure 7. Drift of Common building on the x axis

IOP Conf. Series: Materials Science and Engineering **910** (2020) 012008 doi:10.1088/1757-899X/910/1/012008

Table 4. Comparative analysis of ES / E.030 drifts in the common building

Floor	ES Arequipa 2001	ES Lima 1974	ES Maule 2010	Average
7	1.42	1.04	1.04	1.17
6	1.42	1.04	1.04	1.17
5	1.43	1.04	1.04	1.17
4	1.43	1.04	1.05	1.17
3	1.43	1.04	1.05	1.17
2	1.43	1.04	1.05	1.17
1	1.42	1.04	1.04	1.17
0	1.40	1.02	1.03	1.15

4. Conclusions

The present work has as main objective to estimate the effects of the site spectra of a region, the structural seismic behavior of buildings with two types of importance: an essential and a common one, in addition to making the structural comparison of the seismic response, in this case, mainly drifts, of modal analyzes with design spectra of South American seismic codes, high probability of earthquake enzymes and analysis with site spectra of South American earthquakes, such as Maule, Arequipa and Lima

For the essential building, in relation to the displacements it is concluded that through the modal analysis of the 3 site spectra the structure does not comply with the maximum drifts established in the norm E 030 or the NCh433. Since for Arequipa 2001, which has the maximum representative drift, it is 0.010 on the "x" axis and 0.024 on the y axis. For the common building, it is concluded that the drifts do comply with the provisions of E.030 and in some cases do not comply with NCh433. The representative result was that of Arequipa 2001, although the results of Maule 2010 were very close to those of the previous one mentioned. In "x" the representative drift was 0.0069, which does not comply with the Chilean standard. In "y" the drift was 0.002, which meets both South American standards.

In relation to the essential structure, it can be affirmed that for both S2 type floors the design spectrum of the E030 is above the site spectrum. However, the panorama is different for the common building since the site spectra of Arequipa 2001 represent 142% of S2 in Peruvian regulations. These results follow this trend for the NCH433 standard in the essential building but compared to the E030 the results are much higher in the common since they represent ES Arequipa 2001 for S2 145%. The parameter of the type of use used in both regulations allows to base this situation.

Finally, it is concluded that for the structures analyzed, the effect generated by the site spectra on the essential building is less than that provided in the dynamic modal spectral analysis of Technical Standard E.030 Earthquake-resistant Design and NCH433. However, in multifamily building the answers are much more significant for both standards considering a soil of type S2.

From the previous conclusions it can be affirmed that if it is possible to say that the site spectra are much more significant in an analysis of multifamily dwellings and that the standards of the South American norms would have no problem with an adequate seismic analysis. However, it is recommended for future research, perform the analyzes performed in this work for many more buildings.

5. References

[1] H. Tavera,"Evaluación del peligro asociado a los sismos y efectos secundarios en Perú", IGP, 2014.

[2] A. Eskandarinejad, H. Zafarani and M. Jahanandish, "Local site effect of a clay site in Shiraz based on seismic hazard of Shiraz Plain", Natural Hazards, vol. 90, no. 3, pp. 1115-1135, 2017.

[3] N. Lagaros, C. Mitropoulou and M. Papadrakakis, "Time History Seismic Analysis", Encyclopedia of Earthquake Engineering, pp. 1-19, 2013. Available: 10.1007/978-3-642-36197-5_134-1 [Accessed 20 October 2019].

IOP Conf. Series: Materials Science and Engineering **910** (2020) 012008 doi:10.1088/1757-899X/910/1/012008

[4] Papaspiliou and S. Kontoe, "Sensitivity of site response analysis on the number of ground motion records and implications for PSHA", Bulletin of Earthquake Engineering, vol. 11, no. 5, pp. 1287-1304, 2013. Available: 10.1007/s10518-013-9459-y [Accessed 28 January 2020].

[5] Y. Sonmezer, I. Kalkan, S. Bas and S. Akbas, "Effects of the use of the surface spectrum of a specific region on seismic performances of R/C structures", Natural Hazards, vol. 93, no. 3, pp. 1203-1229, 2018. Available: 10.1007/s11069-018-3347-3 [Accessed 29 January 2020].

[6] Alva, C, & Quispe, E. (2018). Propuesta de procedimiento estandarizado para la obtención de los parámetros de entrada del análisis de respuesta de sitio. Caso de estudio : Provincia constitucional del Callao. Universidad Peruana de Ciencias Aplicadas.

[7] A. Chopra., Dynamics of structures, 4th ed. Boston [etc.]: Pearson, 2014.

IOP Conf. Series: Materials Science and Engineering **910** (2020) 012009 doi:10.1088/1757-899X/910/1/012009

Influence of structural cracking in the seismic response of irregular buildings

K A Álvarez, M F Medina, C A Morales, J D Moreno

Prolongación Primavera 2390, Lima 15023

pccicemo@upc.edu.pe

Abstract- The authors present an investigation on the influence of cracking as a structural response due to a seismic event. The main objective is to determine the impact that cracking has in the seismic response of irregular buildings, and the importance of including inertia-reducing factors in the design of buildings. First, it is proposed to estimate the inertia-reducing factor of the beams, columns and cracked walls from their moment-curvature diagrams. Once those are determined, a comparative analysis will be performed between the seismic responses of regular and irregular buildings of different heights taking into account their gross and effective rigidness. Subsequently, when analyzing the drifts of the buildings with cracked structural elements, a variation of up to 59% with regard to the non-cracked ones. This is a very high percentage that directly influences the considerations indicated in the designing codes. To summarize, it is of high importance to consider the inertia-reducing factors during the structural design process.

1. Introduction

The fundamental principle of international seismic design codes is based on safeguarding the lives of people and their properties against seismic events. By the same token, urban infrastructure must be protected given that it collapses, daily life and rehabilitation after the earthquake becomes difficult task. However, some of the codes do not consider structural in the calculation of seismic responses, which implies a risk element since cracking can occur throughout the building's lifetime, or due to low intensity seismic events.

One of the first people who studied the reduction of stiffness as a result of cracking was Fares [1], who proposed experimental results compared to existing methods, which were only analytical. Finally, he obtained that the estimate of the effective stiffness due to the cracks is within an acceptable range. Moreover, Prestley [2] highlights the importance of considering the cracked sections indicating that, through dynamic analysis to the not cracked structure, it is impossible to find precise seismic responses. Specifically, it indicates that the periods and the distribution of forces of the structure are mostly erroneous.

Li and Xiang [3] analyzed the stiffness reduction factors already established in international standards for structures. From this, they concluded that these factor estimates are not appropriate because they are independent of the reinforcement content and axial load to which the structural elements are subjected, and an equation was proposed to estimate more accurately the effective stiffness.

On the other hand, Castel and François [4] studied total stiffness and irreversible deflection in cracked structural elements. It was demonstrated that the reinforced concrete overall behavior during loading and unloading cycles after cracking represents a significant proportion of the total deflection which the elements were designed for; they also determined that the stiffness of the elements has a progressive decrease according to the cracking level.

In the work carried out by López and Music [5], periods in buildings larger than 10 floors were studied under conditions of cracking and non-cracking in structural walls. The criteria of ACI 318, FEMA 356 and the formula proposed by Paulay and Prestley [6] were used. The results obtained indicate greater displacements between cracked and not cracked buildings in addition to a periods ratio of 1 to 1.5.

Finally, Mohammad Safi and others [7] consider the importance of the effect of cracking in the structural elements, since they directly affect the calculation of natural periods, displacement, among others. For this they used stiffness reduction coefficients proposed in the ACI standard [8] and compared it with the Monte Carlo simulation method. After that, it was shown that the coefficients provided by the standard are not correct since there were variations between 40 -55% in the values of displacements and 20-30% in the period.

In the present study, irregular buildings will be analyzed in order to determine the influence of cracking before seismic applications. A reduction factor of the structural elements will be obtained from the moment-curvature graphs to finally make a comparison of the seismic responses between the cases using the effective stiffness and the gross stiffness.

2. Method

This research evaluates the seismic response in irregular structures with different heights, taking into account the variation in stiffness as a result of cracking, which is observable in the moment-curvature relation of each structural element (columns, beams and plates).

Our proposal assumes already built buildings, however, the method to be used can also be applied to buildings in the design phase. The irregularity to be considered is from incoming corners in the plant. To determine the reduction factors, the moment-curvature diagrams will be used considering the amount of steel, the section of the element, and the constitutive relations of the concrete and its reinforcement.

Figure 1 shows the moment-curvature graph with the most important points to carry out the proposed process, which are the moment of creep and the moment of failure of each structural element; these graphs were obtained computationally.

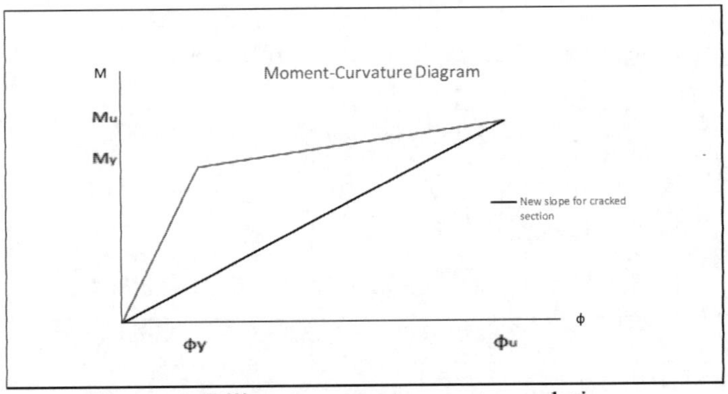

Figure 1. Bilinear moment-curvature relation

The stiffness evaluation, as is known, from the moment-curvature is obtained by:

$$EI = \frac{M_y}{\varphi_Y} \quad (1)$$

Where M_y is the nominal moment that resists the section of the structural element and φ_y is the creep curvature.

Applying [6] for the two critical moments obtained from the graph and comparing its slopes, the reduction factor is determined, with which the cracked inertia is found and then perform the dynamic spectral modal seismic analysis of structures. This analysis generates the seismic responses of regular and irregular buildings, which will be evaluated and compared as a final part of the present investigation.

3. Results

For the validation of the study; 3 regular and 3 irregular buildings of 5, 10 and 15 stories high respectively are proposed. Beams, columns and plates are defined as structural elements, where the beams and columns have 210 kg / cm2 as compressive strength and the plates 280 kg / cm2.
The first model (5 floors) has main beams of section (25x40) cm, (25x20) cm in the secondary beams and (25x40) cm in central and eccentric columns. Similarly, in the second and third models (10 and 15 floors) there are main section beams (35x65) cm, secondary beams (30x50) cm, central and eccentric columns (65x65) cm and 30 cm thick plates.

Table 1 shows the factors obtained for the reduction of stiffness for each structural element.

Table 1. Results of reduction factors.

	Element	Factor
5 Floors	Main beam	0.511
	Secondary Beam	0.546
	Column	0.630
10 Floors	Main beam	0.589
	Secondary Beam	0.578
	Column	0.650
	Plate	0.558
15 Floors	Main beam	0.589
	Secondary Beam	0.578
	Column	0.650
	Plate	0.558

From this, we proceeded to obtain the seismic parameters (periods of vibration, drifts, internal forces) of the 3 buildings considering cracked inertia and gross inertia.

3.1 Periods

The analysis was considered in the first way, since it is where there is greater mass participation. Table 2 shows the periods that exist in each irregular building and how these vary according to their structural cracking and the increase in height of each building.

Table 2. Variation of periods of irregular buildings.

	Not cracked	Cracked	Variation
5 Floors	1.185	1.416	19.49%
10 Floors	0.861	1.087	26.25%
15 Floors	1.303	1.619	24.25%

On average, irregular structures showed a variation due to cracking of 23.33%. On the other hand, to perform the comparison, the analysis was carried out in regular structures, where an average variation of 24.54% was observed. The variation of periods in both regular and irregular buildings, on average, is approximately the same.

3.2 Drifts

IOP Conf. Series: Materials Science and Engineering **910** (2020) 012009 doi:10.1088/1757-899X/910/1/012009

When evaluating the drifts, it was considered to maintain the initial basal shear stipulated by the seismic codes, so that this way the structure is subjected to the same force acting on it. With this, through computational methods, the inelastic drifts of the buildings were obtained to proceed with the comparative analysis

Tables 3 show how much cracking influences when structures undergo an earthquake in the drifts in both directions, demonstrating it by the average variation they represent.

Table 3. Variation of drifts

	Average Variation in Direction X	Average Variation in Direction Y
5 Floors	47.21%	59.00%
10 Floors	43.76%	45.10%
15 Floors	24.35%	34.86%

The irregular structures showed an average variation of 38.44% in the X direction, and 46.32% in the Y direction. Also, the procedure was carried out on regular structures, where an average variation of 34.57% was observed in the X direction, and in the Y direction of 31.01%. Therefore, it can be affirmed that is a greater influence of cracking in buildings with irregularities.

3.3 Internal Forces

The internal forces in the buildings were evaluated to propose more real results to design reinforced concrete structures, considering the variations generated in the results due to cracking.

First, table 4 shows the variations of shear forces on average, both in beams and columns in both directions. It was observed that the ranges of variations reach up to 100%.

Secondly, in table 5 the variation that exists of the moments on average is observed, reaching a maximum of 90%. However, it is also observed that there is a negative variation, which is due to the redistribution of the forces.

Table 4. Variation of shears

	Element	Direction	Variation
5 Floors	Columns	X	30.00 %
		Y	30.00 %
	Beams	X	2.50 %
		Y	5.00 %
10 Floors	Columns	X	50.00 %
		Y	100.00 %
	Beams	X	18.00 %
		Y	60.00 %
15 Floors	Columns	X	32.00 %
		Y	-1.00 %
	Beams	X	8.00 %
		Y	55.00 %

IOP Conf. Series: Materials Science and Engineering **910** (2020) 012009 doi:10.1088/1757-899X/910/1/012009

Table 5. Variation of moments.

	Element	Direction	Variation
5 Floors	Columns	X	27.00 %
		Y	27.00 %
	Beams	X	-5.00 %
		Y	8.00 %
10 Floors	Columns	X	29.00 %
		Y	90.00 %
	Beams	X	15.00 %
		Y	50.00 %
15 Floors	Columns	X	2.50 %
		Y	12.00 %
	Beams	X	20.00 %
		Y	70.00 %

After that, a comparison of the average variations was made with the results obtained in the regular buildings.

Table 6. Comparison of average variations of internal forces in irregular buildings.

	Element	Direction	Variation
Shears	Columns	X	37.00 %
		Y	44.00 %
	Beams	X	9.50 %
		Y	40.00 %
Moments	Columns	X	19.50 %
		Y	43.00 %
	Beams	X	10.00 %
		Y	42.70 %

Table 7. Comparison of average variations of internal forces in regular buildings.

	Element	Direction	Variation
Shears	Columns	X	26.50 %
		Y	5.00 %
	Beams	X	6.50 %
		Y	70.00 %
Moments	Columns	X	9.00 %
		Y	11.50 %
	Beams	X	45.50 %
		Y	61.90 %

Table 6 and 7 show the results obtained where a trend that varies according to the structural element or the internal force analyzed can be observed. Cracking affects more irregular buildings in the case of shear forces, however, in the case of bending moments in beams, it affects more regular buildings.

4. Conclusions

After analyzing the seismic responses of the structures considering the cracking, it has been demonstrated that inertia-reducing factors should be included due to its effects on the design of buildings in order to respond as expected to seismic movements, since drifts increase considerably when cracking is considered. In addition to that, a more conservative design can be obtained if the basal design shear is maintained.

In the same manner, it is observed that the factors of reduction of beams are assimilated to those stipulated by the Complementary Technical Standard of Mexico [9] and the columns and plates resemble the design codes of New Zealand [10], the Complementary Technical Standard of Mexico [9] and the ACI [8]. However, due to the lack of information of some technical standards regarding inertia and cracked stiffness it is recommended to determine the factors taking into account the section, reinforcement, axial load of the element, and its properties.

Finally, if the inertia reduction factors obtained are compared with those recommended by the standards, they are observed to resemble one another. Nonetheless, they are not exactly the same and that is because each designed element is different, so they must have a different factor.

5. References

[1] Fares, N. 1992 Effective Stiffness of Cracked Elastic Solids. *Micromechanical modelling of quasi-brittle materials behavior*. 45 (8), 336-345
[2] Priestley, M. J. N. 2003 Myths and Fallacies in Earthquake Engineering, Revisited. TheMalletMilneLecture. P.P. 9-31
[3] Li, B. y Xiang, W. 2011 Effective Stiffness of Squat Structural Walls. *Journal of Structural Engineering, 131, 1470-479*
[4] Castel, A., François R. 2013. Calculation of the overall stiffness and irreversible deflection of cracked reinforced concrete beams. *Advances in Structural Engineering*. 16 (12), 2035-2042
[5] López, C., Music, J. 2016 Period and displacement analysis of reinforced concrete buildings with different degrees of stiffness in their resistance elements. *Works and Projects,* (9), 33-47.
[6] Paulay, T and Prietsley, M. 1992 "Seismic design of reinforced concrete and masonry buildings". University of California, San Diego, USA: JOHN WILEY & SONS, INC.
[7] Safi, Mohammad and others. 2019. Seismic reliability of concrete buildings considering probabilistic distribution of crack. KSCE journal of civil Engineering. 24 (1), 178-186.
[8] ACI 318 2014 Building Code Requirements for Structural Concrete. Michigan, USA.
[9] NTC-04 2008 Complementary Technical Standards for Design and Construction of Concrete Structures
[10] NZS 3101-06 2006 Concrete Structures Standard.

IOP Conf. Series: Materials Science and Engineering **910** (2020) 012010 doi:10.1088/1757-899X/910/1/012010

Correction coefficients of distortion and vibration period for buildings due to soil-structure interaction

A M Ticona[1], M A Rosales[1], J D Orihuela[1]

[1] Faculty of Civil Engineering, Peruvian University of Applied Sciences San Marcos street 11, Chorrillos, Lima, Peru

Corresponding author's e-mail: u201410718@upc.edu.pe

Abstract. The present research analyzed the influence of the soil structure interaction (SSI) in buildings, varying geotechnical parameters and height, considering 3 international codes. The responses obtained from the structures taking into account the SSI, were compared with the responses of fixed-base buildings, being the main control variables: the period and the drift. It was determined that the estimated range in which the period of the structure increases is from 30 to 98%, demonstrating the influence of considering soil flexibility. Due to the variability of the responses obtained, an adjustment factor is proposed to predict said amplification of the control variables, depending on the height of the building and the ground.

1. Introduction

The conventional analysis performed for the modeling of a building is based on recessing the structure considering an infinitely rigid floor. This idealization does not reflect the real behavior of the soil since it, depending on its geotechnical parameters, provides a certain stiffness that is related to an elasto-plastic behavior [1].

A study prepared by Karapetrou et al. [2] showed that the responses obtained considering the fixed base can lead to non-conservative results such as displacements and periods, which can be amplified in soft soils. Regarding the period, it was determined that the main effect generated by an SSI analysis is the lengthening of this,compared to the rigid base condition, being one of the main causes of this change, the ground in foundation [3]. Similarly, Aydemir et al. [4], evaluated the behavior of structures considering SSI under the effect of an earthquake. The authors verify the increase in the period and demonstrate that the influence of considering such interaction is greater in soft soils. In addition, Joy et al. [5], they proposed a methodology to perform the pushover analysis considering SSI. This study demonstrated the importance of considering the soil-structure interaction in the design of buildings, demonstrating how displacements increase. On the other hand, Valdebenito et al. [6], analyzed the effects of an earthquake of magnitude Mw = 8.2 and the damaged caused to the armed masonry structures. With the explorations carried out, it was demonstrated that the soil-structure interaction played a significant role in the buildings that suffered major damage. The predominant frequency of the soil was very similar to the frequency of buildings with greater structural damage, this correlates with the effects of resonance, an effect that in several of the cases analyzed could have increased seismic demand and thereby justify the increase in damage observed.

This research seeks to determine the impact of considering soil flexibility through an SSI analysis, in relation to two variables: period and drift. The period is one of the most important variables because it allows us to control possible resonances during a seismic event; on the other hand, displacements are closely related to the potential for structural failure. As a main result, it was determined that the percentage increase in the period, for a model that considers SSI, is in a range of 30 to 98% and with respect to drifts between 23 and 145%. Due to the variability in the amplification of the responses obtained, a correlation of data is proposed to establish a function that allows calculating said amplification depending on the floor and the height of the building.

2. Method

To make a correct prediction of the amplification behavior of the control variables, the answers will be calculated considering both idealizations of the soil behavior (SSI and fixed-base). With this, the percentage variation will be determined based on the height of the building, code used and geotechnical parameters for each structure. With the percentages obtained, it will be determined if the correlation of data is correct. For this, the proposed function will be validated by checking with a 95% reliability. This reliability will be applied to the own coefficients of the established regression.

The use of the equations, for the calculation of the amplification factors, is limited to a foundation system consisting of combined footings, for structures that vary between 4 floors (12 meters) to 12 floors (35 meters) and soil types employees in the study.

3. Analysis

In the first instance we will proceed with the analysis considering fixed-base. The buildings will be modeled and analyzed with the ETABS 2016 computational tool. The resulting drifts for both directions are shown in the graphs (figure 1 and figure 2), varying the soil parameters and the height of the building.

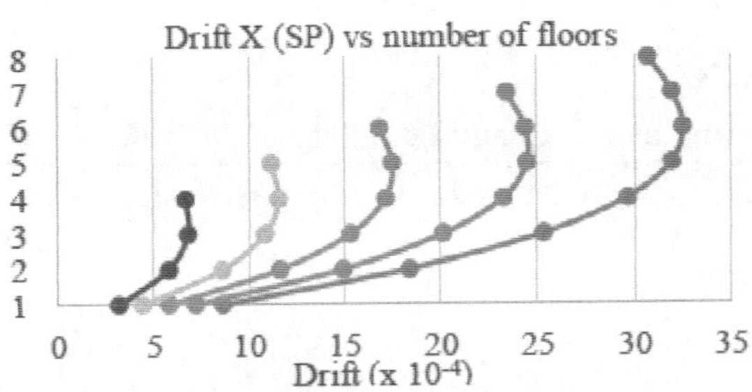

Figure 1. Drift X (SP) from 4 to 8 floors

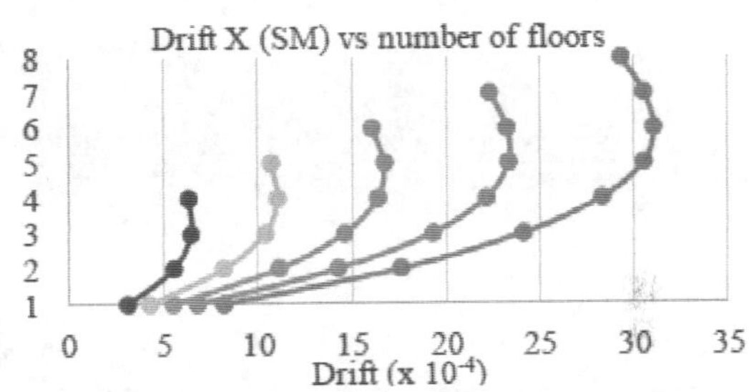

Figure 2. Drift X (SM) from 4 to 8 floors

After verifying the structure, the SSI analysis is carried out. For this, it is necessary to calculate the ballast coefficient that allows us to find an equivalence of the soil, similar to the behavior of a spring. For this evaluation of the structure, it was designed considering 3 codes: that of the American code ASCE 41-13 [7], Russian code SNIP 2-02-05-87 [8] and Mexican complementary technical code [9]. In addition, data on the soil parameters that are required by such design regulations is necessary (table 1).

Table 1. Geotechnical parameters of silty and poorly graded sand

Soil parameters	SP (Poorly Graduated Sand)	SM (Silty Sand)
Soil elasticity modulus (kg/cm^2)	190	200
Poisson's ratio	0.25	0.22
Internal friction angle (°)	26.7	27
Volumetric density (gr/cm^3)	1.681	1.52
Spread Wave Propagation Speed (m/s)	180	250

Then we proceed to calculate the 6 ballast coefficients (in the direction X, Y, Z and their respective turns). Subsequently, the support of the structure is released from the embedment to assign said coefficients, which will idealize the behavior of the land. With the help of the software, the deformation of the mezzanines and the period of the structure are calculated considering the flexibility of the floor.

As evidence is shown the results obtained from the American code ASCE 41-13 for an SP soil (table 2 to 3):

Table 2. Drift in X considering SSI according to American code ASCE 41-13 for a soil SP.

N° of floors	Drift Y - American code ASCE 41-13				
8	0.0051				
7	0.0052	0.0041			
6	0.0054	0.0042	0.0031		
5	0.0054	0.0043	0.0032	0.0023	
4	0.0053	0.0043	0.0033	0.0023	0.0015
3	0.0050	0.0041	0.0032	0.0023	0.0016
2	0.0045	0.0037	0.0029	0.0022	0.0015
1	0.0043	0.0036	0.0029	0.0023	0.0017

Table 3. Drift in X considering SSI according to American code ASCE 41-13 for a soil SP.

N° of floors	Drift X - American code ASCE 41-13				
8	0.0044				
7	0.0046	0.0035			
6	0.0047	0.0037	0.0027		
5	0.0047	0.0037	0.0028	0.0019	
4	0.0046	0.0037	0.0028	0.0020	0.0013
3	0.0043	0.0035	0.0027	0.0020	0.0013
2	0.0038	0.0031	0.0025	0.0019	0.0013
1	0.0036	0.0030	0.0025	0.0019	0.0014

Table 4. Period (seconds) of the structure considering recessing and flexibility in the base.

	Fixed-base	SSI - Russian Code		SSI – American Code		SSI – Mexican Code	
N° floors	SP o SM	SP	SM	SP	SM	SP	SM
8	0.43	0.652	0.645	0.58	0.515	0.623	0.618
7	0.348	0.551	0.545	0.486	0.427	0.526	0.522
6	0.272	0.455	0.449	0.396	0.344	0.434	0.43
5	0.203	0.363	0.358	0.312	0.266	0.348	0.344
4	0.141	0.277	0.273	0.234	0.195	0.267	0.264

As can be seen in the table 4; there is an important variation in the responses when considering the soil as an elasto-plastic element. The present investigation seeks to propose a correlation between the responses considering SSI and those of an analysis fixed in the base, with the aim of establishing a correction parameter for both the period and the mezzanine drift.

4. Results

After the analysis of the structure considering the flexibility of the soil and fixed in the base, we proceed to determine the regression that best fits the data obtained. The data that will be used for this study will have the following characteristics: the first 3 floors of the analyzed structures will not be used, since the variation of their results (SSI vs fixed), does not fit the data dispersion and the correlation will be established for structures greater than 12.00m (or 4 floors) at 35.00m (or 12 floors).

4.1. Data correlation analysis for the drifts obtained

In the first instance, correlations will be established for the variation of the drifts according to the height of the structure (Cd as a function of Ht). The variable Ht defines the building height in meters and the variable Cd defines the percentage or coefficient of drift amplification considering ISE vs fixed-base building. Consequently, the correlation is established based on the height of the building, code used and type of soil. These results can be seen in table 5, 6 and 7.

Table 5. Model and correlation factor that was designated for the Russian code SNIP 2-02-05-87 for SP and SM soil

	Russian code (SP)	Russian code (SM)
Correlation Model	Logarithm of X	Logarithm of X
Correlation	75.79%	81.62%

Table 6. Model and correlation factor that was designated for the American code ASCE 41-13 for SP and SM soil

	American code (SP)	American code (SM)
Correlation Model	Logarithm of X	Logarithm of X
Correlation	81.50%	87.13%

Table 7. Model and correlation factor that was designated for the Mexican complementary technical code for SP and SM soil

	Mexican code (SP)	Mexican code (SM)
Correlation Model	Logarithm of X	Logarithm of X
Correlation	91.71%	93.11%

These correlations were chosen because these are the ones that best fit the dispersion of data (Cd depending on the height of the building). Subsequently, the regression of data for each correlation is validated. It should be mentioned that the amount of data used for the statistical analysis is 30 values for each regression. It was obtained that the P-value (probability corresponding to the statistic if possible, under the null hypothesis) is less than 0.05; therefore, there is a statistically significant relationship with a 95.0% confidence level in all cases [10]. The functions established for the calculation of Cd as a function of Ht are the following:

$$Cd_{RC\text{-}SP}=3.9104-1.0075\ \ln(H_t) \quad (1)$$
$$Cd_{MC\text{-}SP}=3.5146-0.9071\ \ln(H_t) \quad (2)$$
$$Cd_{AC\text{-}SP}=4.6232-1.1847\ \ln(H_t) \quad (3)$$
$$Cd_{RC\text{-}SM}=4.0671-1.078\ \ln(H_t) \quad (4)$$
$$Cd_{MC\text{-}SM}=3.531-0.9226\ \ln(H_t) \quad (5)$$
$$Cd_{AC\text{-}SM}=2.6233-0.6785\ \ln(H_t) \quad (6)$$

As it was observed in the regression of the data; the mexican code has a higher correlation of data, so it would fit a better prediction of drift amplification based on their responses. However, the american code is the one that generates the greatest amplification in drifts (for the SP soil), although these results are significantly different from those obtained through the Russian and Mexican code. In figure 3 the variation of the functions defined for each code can be observed.

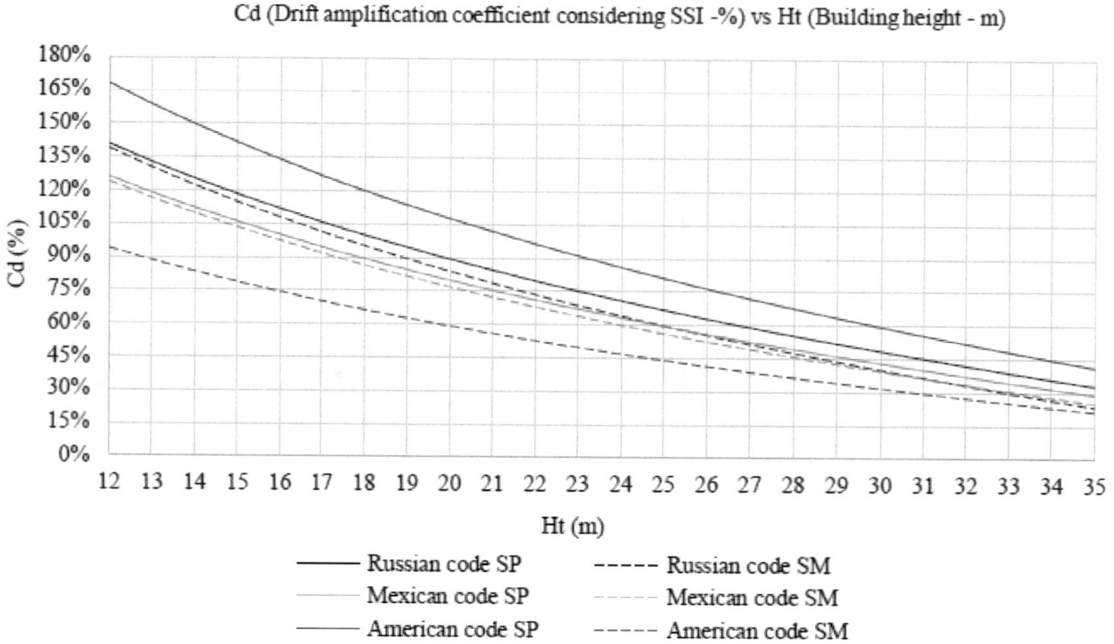

Figure 3. Cd depending on the height of the building for the Russian, Mexican and American code considering the soil SP and SM

Due to the variability of the regressions obtained from the 3 models studied, an equation is proposed that will more optimally normalize the distribution of the data to predict the value of the amplification coefficient Cd. These proposed equations have a correlation of 74.04% and 75.91% respectively, with a P-value less than 0.05 (figure 4).

$$Cd=4.0161-1.0331\ \ln(H_t) \text{ to soil SP} \quad (7)$$

$$Cd=3.4071-0.893\ \ln(H_t) \text{ to soil SM} \quad (8)$$

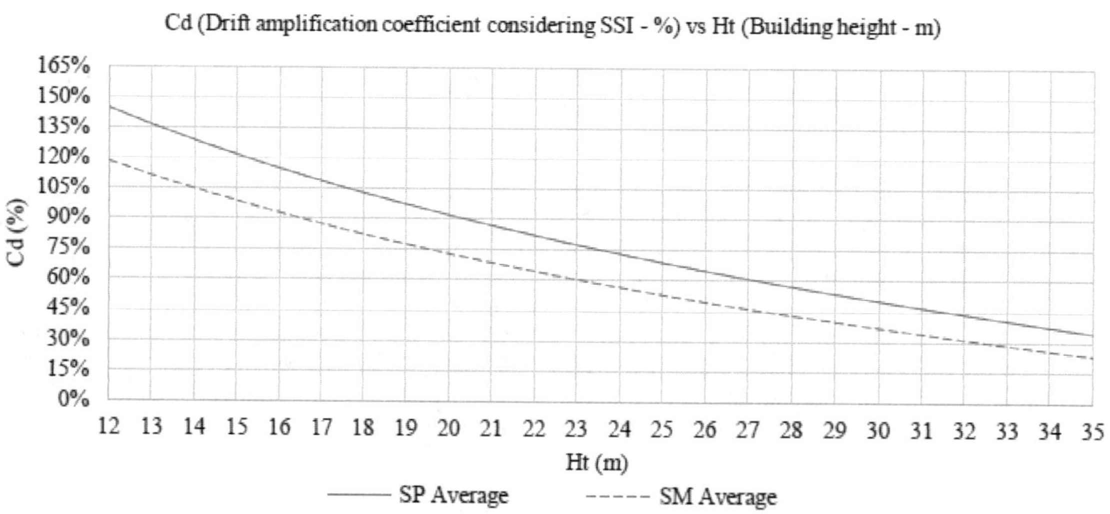

Figure 4. Cd depending on the height of the building considering the soils SP and SM

4.2. Data correlation analysis for periods

Therefore, the regression will be carried out for the periods considering the variables Ct and Ht. It should be noted that the correlation of data regarding the period of the structure exceeds 99% with a P-value of less than 0.05, so all regressions are apt to predict the behavior of the period. The variable Ht defines the building height in meters and the variable Ct defines percentage or coefficient of period amplification considering ISE vs fixed-base building (figure 5)

The functions established for the calculation of Ct based on Ht considering each code are the following:

$$Ct_{RC\text{-}SP}=(0.1412+0.0749\ Ht)^{-1} \quad (9)$$
$$Ct_{MC\text{-}SP}=(0.1714+0.0644\ Ht)^{-1} \quad (10)$$
$$Ct_{AC\text{-}SP}=\exp(2.3558-0.9938\ \ln(H_t)) \quad (11)$$
$$Ct_{RC\text{-}SM}=(0.144+0.0773\ Ht)^{-1} \quad (12)$$
$$Ct_{MC\text{-}SM}=(0.1906+0.1146\ Ht)^{-1} \quad (13)$$
$$Ct_{AC\text{-}SM}=\exp(2.3291-0.9933\ \ln(H_t)) \quad (14)$$

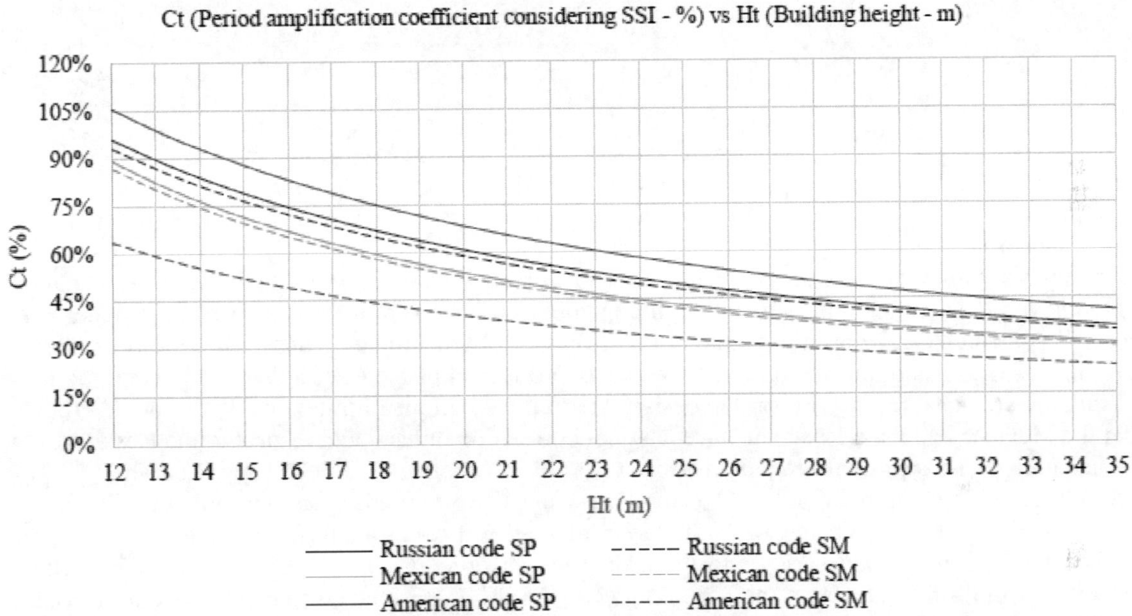

Figure 5. Ct depending on the height of the building for the Russian, Mexican and American code considering the soil SP and SM

The regression of data in terms of periods is very similar to the behavior of the drifts obtained, so a coefficient that represents the behavior of Ct is also proposed. The following equations have a correlation of 93.25% and 82.70% respectively with a P-value less than 0.05 (figure 6).

$$Ct=0.061+10.9705\ Ht^{-1} \quad (15)$$
$$Ct=0.0387+9.3288\ Ht^{-1} \quad (16)$$

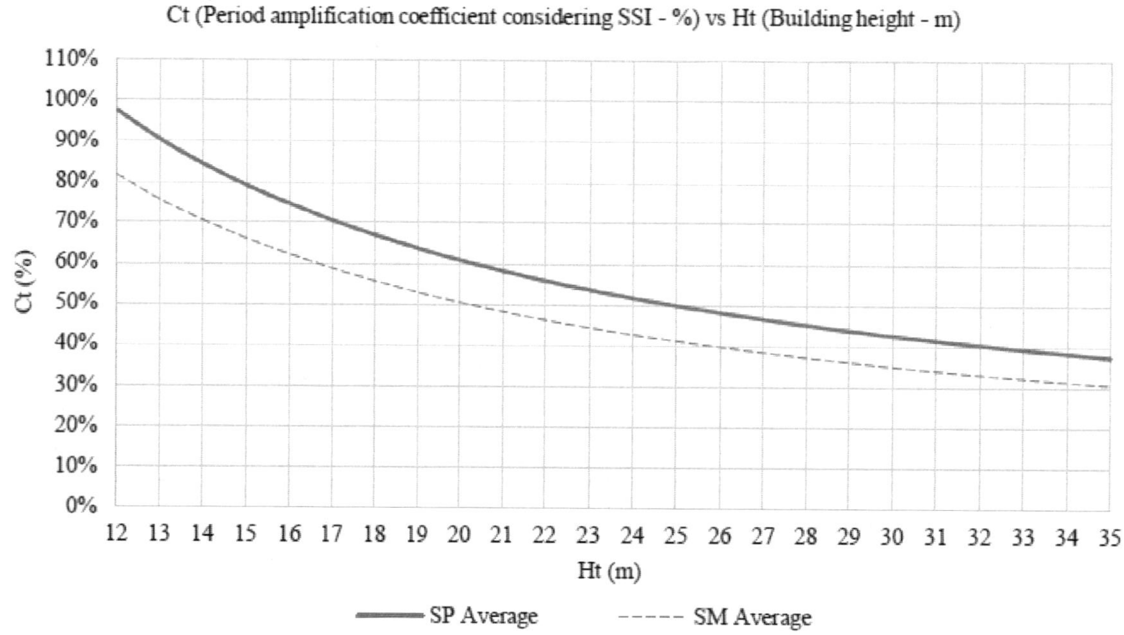

Figure 6. Ct depending on the height of the building considering the soils SP and SM

5. Conclusions

In the present work, we have proposed functions with a statistically significant relationship that allow us to know the period and the drifts of a building considering SSI from the responses of a model fixed in the base, which can be used to predict a real behavior of the structure considering the flexibility of the soil. Its application must be used for a preventive criterion of the effects that can generate failures or collapses (resonance) but not for the design due to the mentioned limitations. It can be determined that the Mexican and American code have a higher correlation of data with respect to the Russian code, the percentage of correlation being for Mexican C. 93.41%, American C. 84.32% and Russian C. 78.71% with respect to the amplification of the drifts However, it is proposed to use the functions Ct and Cd, which together predict the behavior of the three international codes analyzed. On the other hand, the method that considers values closer to the proposed functions (Ct and Cd) is the Mexican method, however, if the soil is very soft, the American code ASCE 41-13 method must also be considered to respect the maximum limit of drifts provided by any of the three code mentioned.

It is recommended to develop studies for the elaboration of a multiple regression model that considers the lateral stiffness and the height of the structure, which would modify the equations proposed in the present investigation.

6. References

[1] Barkan DD. Dynamics of bases and foundations. New York: McGraw-Hill, 1962, pp. 1-51.

[2] Karapetrou ST, Fotopoulou SD, Pitilakis KD. Seismic vulnerability assessment of high-rise non-ductile RC buildings considering soil-structure interaction effects. Soil Dyn Earthq Eng [Internet]. 2015; 73:42–57. http://dx.doi.org/10.1016/j.soildyn.2015.02.016.

[3] Pérez-Rocha LE, Avilés J. Damage Analysis in Structures with Flexible Support. Seismic engineering magazine. 2007; 77:89–111. https://doi.org/10.18867/ris.77.101

[4] Aydemir ME, Ekiz I. Soil-structure interaction effects on seismic behaviour of multistorey structures. Eur J Environ Civ Eng. 2013;17(8):635–53. https://doi.org/10.1080/19648189.2013.810177.

[5] Joy PV, Kuriakose B, Mathew M. Pushover Analysis of Buildings Considering Soil-StructureInteraction. Appl Mech Mater. 2016; 857:189–94. https://doi.org/10.4028/www.scientific.net/AMM.857.189

IOP Conf. Series: Materials Science and Engineering **910** (2020) 012010 doi:10.1088/1757-899X/910/1/012010

[6] Valdebenito G, Alvarado D, Sandoval C, Aguilar Vidal V. Iquique earthquake Mw = 8.2 – 01 abril 2014: Observed damage and site effects in masonry structures. 2015.

[7] Seismic evaluation and retro adaptation of existing buildings: ASCE/SEI 41-13 Recovered。 https://mega.nz/#!zFhhGKTI!bNsBkUruCUt5gyDszpXLSR9HSCoph4EhId91jFt7n7U [consultation: November 21, 2019].

[8] Construction rules and regulations 2.02.05-87. Design of foundations for dynamic loads. Recovered from https://files.stroyinf.ru/Data2/1/4294854/4294854681.pdf [consultation: November 21, 2019].

[9] Complementary technical standards for earthquake design. Recovered from http://www.smie.org.mx/layout/normas-tecnicas-complementarias/normas-tecnicas-complementarias-diseno-sismo-2017.pdf?fbclid=IwAR0XHZmzb_kwF1o-Rs8rOhP7QvRGIBvOhZbUutDo8cEDGhwEcAHuzmgL8_4 [consultation: November 21, 2019].

[10] Gardner MG, Altman DG. Confidence intervals instead of P values: hypothesis test estimate. BMJ.1986; 292. https://dx.doi.org/10.1136%2Fbmj.292.6522.746

A comparison of diurnal variation of pavement albedo between vertical and horizontal surfaces under tropical climatic condition of Thailand

M Srivanit[1,2,3] and D Jareemit[1,2]

[1] Faculty of Architecture and Planning, Thammasat University, Pathumthani, Thailand.
[2] Research Cluster in Livable Environment and Architectural Design Laboratory (LEAD-Lab)
[3] Center of Excellent in Urban Mobility Research and Innovation (CoE-UMRI)

E-mail: s.manat@gmail.com

Abstract. Albedo is an important indicator of radiation reflectance of pavement surfaces for the building envelope and on the ground level, and their resultant impacts on human comfort and the urban environment in outdoor spaces. Usually, albedo is generally accepted only for horizontal surfaces. This study developed an experimental test set-up for the albedo measurement system with pyranometers and automatic meteorological data acquisition system, which used it to conduct field measurements of albedo on horizontal and vertical surfaces of the used concrete block. The results show that albedo measured on a horizontal surface is not proportional to irradiance on a vertical surface. The albedo value between the two surfaces depend upon time of day, and the horizontal surface also received significantly more incident solar radiation than the vertical surface during all but the central hours of the day, while at reflected solar radiation on vertical irradiances were less than the horizontal. These results can help reduce the uncertainty in understanding and evaluating the thermal behaviour of the building and environmental impacts of pavement surfaces with different albedos in the outdoor urban space.

1. Introduction

The most climatic condition of Thailand is tropical savanna climate that corresponds to the Köppen climate classification categories "*Aw*" in the northern hemisphere [1]. Moreover, a large percentage of Thailand's urban areas is covered with a high density of pavement and increasing in the pavement surfaces, which are contribute to significantly creation of the surface urban heat island (UHI) by solar radiation [2,3]. When the pavement is exposed to the solar energy emitted by the sun ray, some of total solar radiation on pavement surface is reflected back into the air, and some is absorbed by the pavement, which will lead to temperature rise in the surface pavement and higher temperatures of urban area than that of rural area [4,5]. In the northern hemisphere, the winter sun (November, December, January) rises in the southeast, transits the celestial meridian at a low angle in the south (more than 43° above the southern horizon in the tropics), and then sets in the southwest. A vertical surface facing south of building is effective for capturing solar thermal energy [6], and the sun ray is a major factor in the heat gain of buildings which can create thermal effects on urban environment.

Thus, albedo (or solar reflectivity) is an indicator of the reflecting power of a surface and a key thermal characteristic, which defined as the ratio of the reflected solar radiation to the incident solar radiation at the surface [7], and is generally accepted only for horizontal surfaces [8]. An albedo of 0 means no reflecting power of a perfectly black surface (none reflected, all absorbed), an albedo of 1 means perfect reflection off a perfectly white surface [9]. Incident irradiance on a non-horizontal surface (from any surface position of vertical or inclined) from a variety of incident angles may cause the reflectivity can change. Assumptions about the reflectivity of a vertical surface are frequently made for a variety of wall surfaces usually found in urban buildings. In this work, the relation between albedos on a horizontal surface, and that on vertical surface facing directly south of the used concrete blocks (the whole measurement system is shown in figure 1), has been investigated under ideal conditions throughout a single day with a diurnal range of zenith angles and also to examine diurnal changes in albedo.

2. Research Methodology

Reflectance measurements on vertical and horizontal surfaces was measured throughout a cloudless day (24 January 2020, as in the winter time in the northern hemisphere) at 7th floor rooftop of the Faculty of Architecture and Planning building at Thammasat University (Rangsit Campus) (14.06° N, 100.61° E), using a dual-pyranometer is composed of two S-LIB-M003 silicon pyranometer sensors with a measurement range of 0 to 1280 W/m^2 over a spectral range of 300 to 1100 nm and HOBO automatic meteorological data acquisition system (as shown in figure 1(c)), and measure distance of 40 cm from base of the sensor to sample. Two types of concrete blocks were applied on two-frame substrate of wood comprises a horizontal and vertical frames with 1.2x1.2 m, which is the reference material for comparison (figure 1(a), 1(b)). The climatic data (including air temperature, relative humidity, wind speed and direction) during the measurement period were also monitored using a nearby HOBO weather station. During the measurement day of January 24th 2020, temperature daily highs to be around 35 °C with humidity averages just under 90%, prevailing wind are mostly south direction. The climatic conditions on an experimental day in this study is summarized in figure 2(a). To ensure that the measurement location had the efficiency surface solar radiation under clear sky condition as that obtained by using the 180° fisheye lens to take a test point image. This study calculated the sky view factor (SVF) value in RayMan Software, as the percentage of level of sky clearness at a measurement point. As shown in figure 2(b), it is expressed as a 9.2% value of the horizon limitation, representing the measurement location located at the relatively open area. The sun path diagram shows the sun rise in the southeast and sets in the southwest. At noon, it looms in the middle of the horizon and directly south.

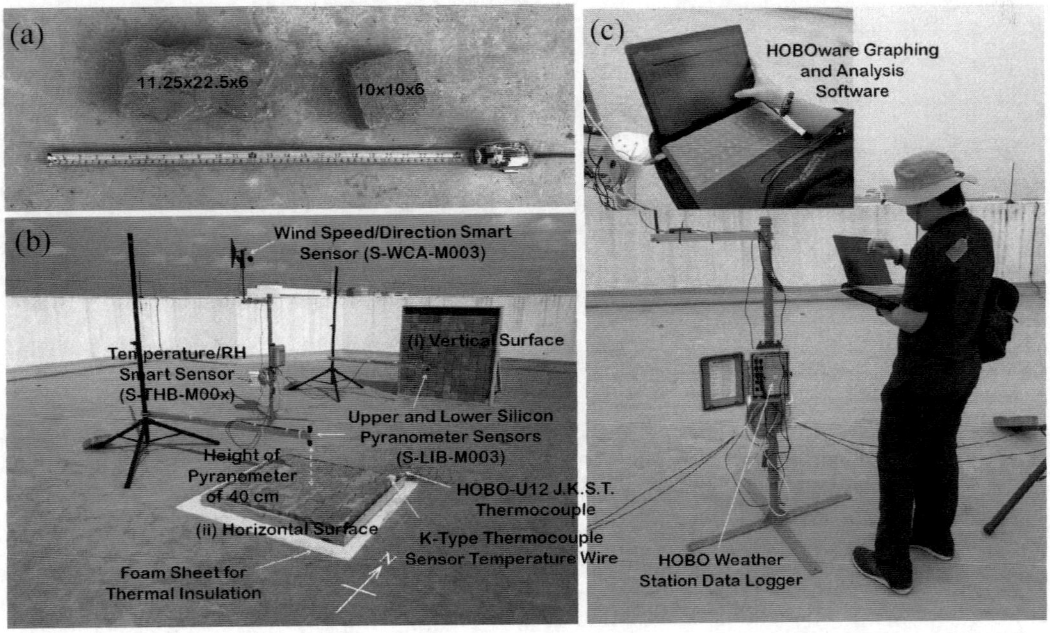

Figure 1. Illustration of (a) two types of used concrete blocks with different size, and (b) installation of a dual-pyranometer to measure incoming and reflected solar irradiance of pavement, (c) which can be recorded together automatically using a data logger.

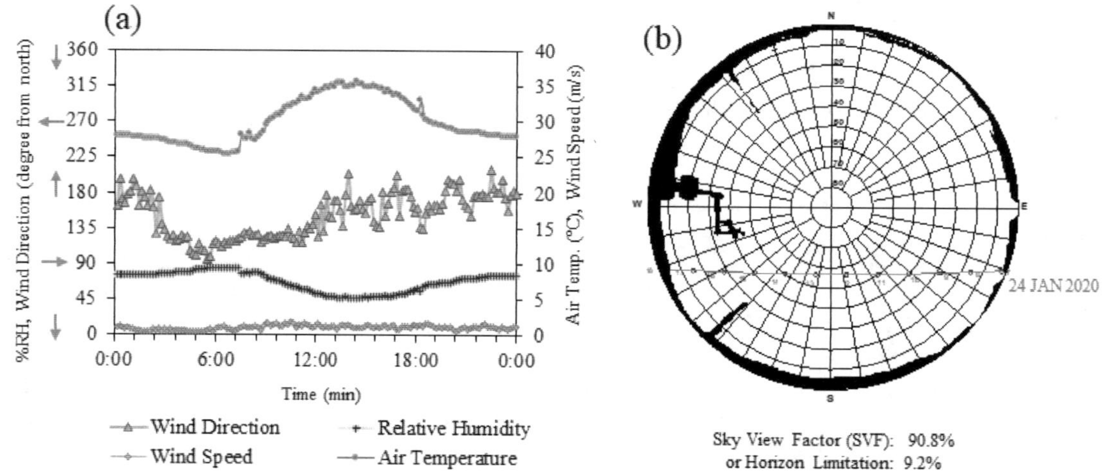

Figure 2. (a) The climatic conditions on an experimental day, and (b) sun path diagram and the SVF percentage of the measurement location in January 24th, 2020.

3. Results and Discussion

3.1. Diurnal-albedo due to surface types

The measured albedos are illustrated in figure 3 on all surfaces using a boxplot, gives an even clearer illustration of this variation of albedo across these two surfaces. The results show that a horizontal surface, which is facing directly south, has lower albedo than a vertical surface with the diurnal-mean albedo are 0.43 and 0.45, respectively. The relatively low albedo of horizontal surface will absorb more incident solar radiation, and produce a high temperature. In contrast, a vertical surface generally has a higher albedo, which will reflect more incident solar radiation and produce a lower pavement temperature.

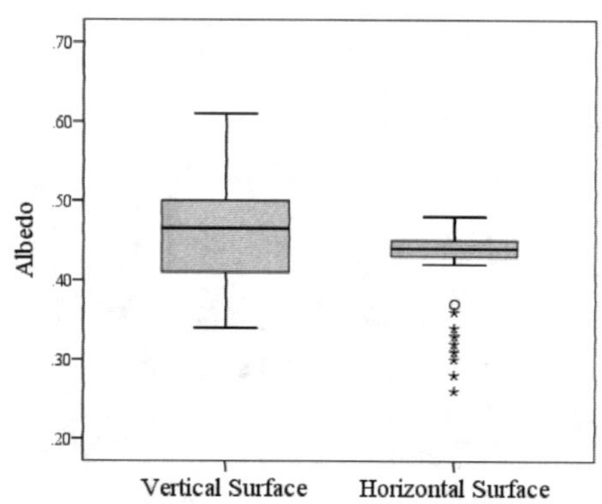

Statistics of diurnal-albedo	South facing vertical surface	Horizontal surface
Mean	0.45	0.43
SD	0.06	0.05
Median	0.46	0.44
Min	0.34	0.26
Max	0.61	0.48
Max-Min	0.28	0.22

Figure 3. Boxplot and summary statistics showing the surface variations of diurnal-albedo of concrete blocks measured on January 24th, 2020.

3.2. *Effect of solar radiation intensity on diurnal-albedo*

The incident and reflected radiation of the concrete block pavement surfaces were measured respectively on January 24th, 2020. The results of radiation at every ten minute were shown in figure 4. It can be clearly seen that the incident solar radiation intensity fluctuated violently even within ten minutes on both vertical and horizontal surfaces. As expected, the incident solar radiation at the vertical surface (calculated between the local sunrise (6:45 am) and sunset (6:19 pm) times) ranging from 13 W/m^2 to 568 W/m^2 is much lower than that at the horizontal surface ranging from 24 W/m^2 to 852 W/m^2. However, the reflected radiation measured at the vertical surface is in the range of 9 W/m^2 to 244 W/m^2, which is close to the reflected radiation measured at the horizontal surface ranging from 8 W/m^2 to 368 W/m^2. This implies that the incident angle has effect on the measured albedo. In the case of horizontal surface is plotted in figure 4(b), a higher albedo will be given on the early morning or late afternoon, and relatively constant around the middle of the day (11:00 am to 3:00 pm).

Measured albedo changes over time during day on a south facing vertical surface, it is high in the early morning and then lead to a continuous decline in the late afternoon when there is a low incident angle of solar radiation as in the early morning (see figure 4(a)). The position of the early morning sun means that direct solar radiation is positioned to fall more directly on the south facing vertical surface, causing maximum reflection of the incident solar radiation to the detector for a south facing orientation. The decrease in diurnal-albedo value is due to the change in sun azimuth from morning to afternoon, where the direct solar radiation falling on the vertical surface has decreased due to the westerly progression of the sun through the sky. This was also observed by Webb et al. [10] and Turner et al. [8] when investigating irradiances falling on vertical surfaces in the southern hemisphere.

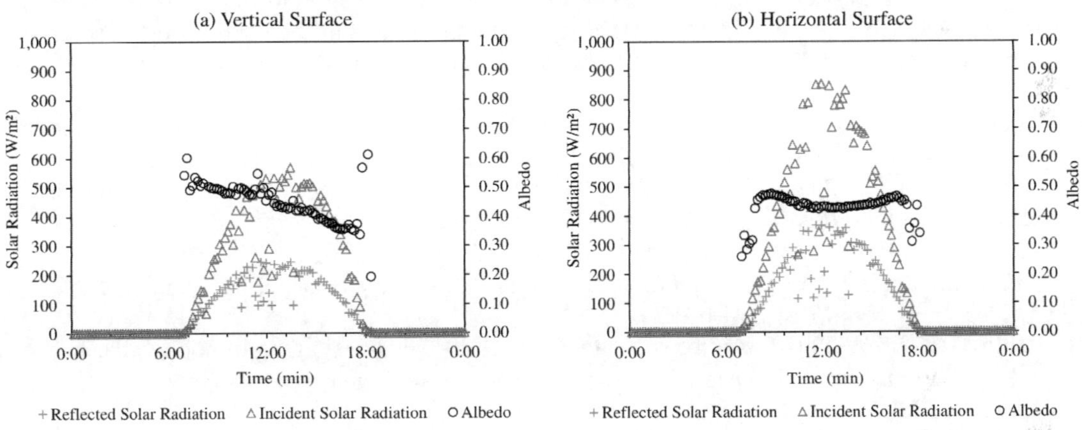

Figure 4. Diurnal variation of solar reflectivity on concrete blocks in January 24th, 2020 for (a) south facing vertical and (b) horizontal surfaces.

4. Conclusion

An experimental test set-up for the albedo measurements presented in this study are for a single day in winter time in the northern hemisphere at Thailand. The diurnal variation of albedo determined from the dual-pyranometer measurements on vertical and horizontal surfaces is analysed. The results illustrate the changing nature of albedo on a south facing vertical surface compared to that of a horizontal surface. The study found that the sun have higher incident solar radiation at large albedo on a horizontal surface of concrete blocks than a south facing vertical surface. The south facing receives comparatively little radiation at Thailand location. It may be because when the position of the sun is to the south it is also very close to the zenith position (the winter month of January), thus a vertical surface is far from normal to the direct beam [11,12]. These results take on a stronger connotation if extended to other seasons of Thailand. Additionally, a further study will be to perform a sensitivity analysis of the albedo enhancement of selected surfaces to reveal a more accurate assessment of urban climate of Thailand.

The measurement system in this study can be applied quantitatively to other pavement surfacing materials and conditions, but by studying what happens in these ideal conditions a more general qualitative pattern of albedos can be inferred and applied to other situations [13,14]. However, additional simulations need to be carried out to investigate the combined effects of simultaneously increasing albedo of pavements and other building façade materials on urban climate and building energy use, and their impacts on human thermal comfort [15,16,17,18].

5. References

[1] Kottek M, Grieser J, Beck C, Rudof B and Rubel F 2006 World map of the Köppen-Geiger climate classification updated *Meteorologische Zeitschrift* 15 (3) 259-263
[2] Oke TR 1982 The energetic basis of the urban heat island *QJRMS* 108 1-24
[3] Srivanit M, Hokao K and Phonekeo V 2012 Assessing the Impact of urbanization on urban thermal environment: a case study of Bangkok metropolitan *Int. J. Appl. Sci. Technol.* 2 (7) 243-256
[4] Bretz S, Akbari H, Rosenfeld AH and Taha H 1992 Implementation of solar reflective surfaces: materials and utility programs *University of California Berkeley* LBL Report 32467
[5] Doulos L, Santamouris M and Livada I 2004 Passive cooling of outdoor urban spaces. The role of materials *Sol. Energy* 77 231-249
[6] Kalogirou SA 2014 Chapter 2 - Environmental Characteristics *Solar Energy Engineering (Second Edition)* ed Boston: Academic Press 51-123
[7] Li H, Harvey J and Kendall A 2013 Field measurement of albedo for different land cover materials and effects on thermal performance *Building and Environment* 59 536-546
[8] Turner J, Parisi AV and Turnbull DJ 2008 Reflected solar radiation from horizontal, vertical and inclined surfaces: Ultraviolet and visible spectral and broadband behaviour due to solar zenith angle, orientation and surface type *Journal of Photochemistry and Photobiology B: Biology* 92 (1) 29-37
[9] Chen J, Zhou Z, Wu J, Hou S and Liu M 2019 Field and laboratory measurement of albedo and heat transfer for pavement materials *Construction and Building Materials* 202 46–57
[10] Webb AR, Weihs P and Blumthaler M 1999 Spectral UV irradiance on vertical surfaces: a case study *Photochem. Photobiol.* 69(4) 464-470
[11] Exell RHB 1976 The solar radiation climate of Thailand *Solar Energy* 18(4) 349-354
[12] Hiranvarodom S 2005 Estimation of solar radiation on inclined surfaces for Thailand *Journal of Engineering. RMUTT* 3(6) 101-107
[13] Paolini R, Zani A, Poli T, Antretter F and Zinzi M 2017 Natural aging of cool walls: Impact on solar reflectance, sensitivity tothermal shocks and building energy needs *Energy and Buildings* 153 287–296
[14] Paolini R, Terraneo G, Ferrari C, Sleiman M, Muscio A, Metrangolo P, Poli T, Destaillats H, Zinzi M and Levinson R 2020 Effects of soiling and weathering on the albedo of building envelope materials: Lessons learned from natural exposure in two European cities and tuning of a laboratory simulation practice *Solar Energy Materials & Solar Cells* 205 110264
[15] Erell E, Pearlmutter D, Boneh D, Kutiel PB 2014 Effect of high-albedo materials on pedestrian heat stress in urban street canyons *Urban Climate* 10 367-386
[16] Schrijvers PJC, Jonker HJJ, de Roode SR and Kenjereš S 2016 The effect of using a high-albedo material on the Universal Temperature Climate Index within a street canyon *Urban Climate* 17 284-303
[17] Taleghani M 2018 The impact of increasing urban surface albedo on outdoor summer thermal comfort within a university campus *Urban Climate* 24 175-184
[18] Jandaghian Z and Berardi U 2020 Analysis of the cooling effects of higher albedo surfaces during heat waves coupling the Weather Research and Forecasting model with building energy models *Energy and Buildings* 207 109627

Acknowledgments
This study was supported by Thammasat University Research Fund, Contact No. TUGT 2/2562. We especially want to thanks Nichapat Phlaiphlerd and the team of the LEAD-Lab for assisting this research project.

Risk-assessment of hydropower plant susceptible to seismic hazard by 3D spectrum analysis

L Roslan[1,*], R Ahmadi[1], K K F Lau[1], L T Trinh[2] and M Rosyadi[3]

[1] Universiti Malaysia Sarawak, 94300 Kota Samarahan, Sarawak, Malaysia
[2] Asian Institute of Technology, 58 Moo 9, Km. 42, Paholyothin Highway Klong Luang, Phatum Thani, 12120 Thailand
[3] PT Indah Karya, Jalan Golf No. 2A Arcamanik Bandung, 40294, Indonesia

Corresponding author: rlidyana@unimas.my

Abstract. Recently, new fault zones are identified in central Borneo due to the ongoing geomorphological anomalies along the Rajang-Crocker Belt. The Kapit-Belaga region along the Rajang River system are located in between active trending strike-slip faults. By 2025, this region will have 3 operational large hydropower dams namely, Bakun, Murum and Baleh which are approximately 120 km apart in radius. Even though central Borneo is known to be a stable area, the assessment of seismic risk for existing dams is vital. The seismic failure of Konya Dam, India in 1967 proved that occurrences of catastrophic earthquake in low seismicity region is possible. Furthermore, earthquake frequency model shows that it is plausible for the Crocker fault system to produce seismic ruptures of magnitudes $\geq$7.0. The research of risk-assessment for hydropower in this work is motivated by the possibility of larger seismic hazard within the Crocker fault system. Hypothetical seismic loadings from Tosya and Aydin earthquakes' records are adopted as inputs to replicate long-distance waves for the Response Spectrum Analysis (RSA) by ANSYS© ACADEMIC. Results show that tensile zone develops at the reservoir bed, extending to the dam foundation. When the seismic load is at 0.625 g Peak Ground Acceleration (PGA) – which is equivalent to magnitude 6.0 earthquake, the mechanical responses of the dam body excitation and hydrodynamic pressure from the impact of the reservoir due to seismic excitation triggers structure failure at the mid crest and bottom of the dam.

1. Introduction

Borneo is a continental region which elastic thickness of the lithosphere is thin, with high heat flow and low velocities of sediment yielding at shallow depths in the mantle [1]. Due to the ongoing geomorphological anomalies along the Rajang-Crocker Belt [2], new dextral strike-slips which deflected streams and sinistral faults appears, surrounding central Borneo [3]. The Kapit-Belaga region along the Rajang river system are located in between these major strike-slip faults. By year 2025, Sarawak will have 3 operating large dams in 120 km radius of the Kapit-Belaga region, namely Bakun Murum and Baleh. Bakun and Murum are built on fault lines underlying the Rajang River System. Although Borneo is relatively stable, strong seismic motion were recorded by the Malaysia Meteorological Department's seismological station (Station KASM) near the region of responsible [4]. The Konya, India earthquake in 1967 is an example that earthquakes with surface-wave magnitude of 6.5 can occur in a region considered stable with no apparent seismicity [5]. Thus, the risks of large earthquake occurrences in the Kapit-Belaga region cannot be dismissed. Furthermore, earthquake frequency model suggests the reoccurrences of large earthquake caused by the Crocker fault system after the 2015 Ranau earthquake in Sabah, Malaysia is plausible [6]. On top of that, the peak ground acceleration (PGA) for 500 years return period in Malaysia has doubled from 0.5 g PGA to 1.0 g PGA after the Aceh earthquake in 2004 [7]. Due to the characteristics of the tectonic settings in Borneo[1-3], the low rates of average fault slips mean that earthquake disasters with return period measured in thousands of years can occur unexpectedly because they are difficult to estimate [8]. Since earthquake with destructive characteristic in the Asia region is caused by the long period vibration [7] and long-wave propagated from various epicentres [9], there are risks for Bakun dam or Murum dam to endure seismic excitation caused by strike-slip rupture around the region.

Therefore, to mitigate seismic hazard of operating dams, it is crucial to assess the seismic risks. Our previous studies analysed the hypothetical risk assessment of Murum dam when introduced to long long-wave propagation [10]. The results showed that Murum dam can withstand up to 0.1 g PGA shaking and the critical cracking occurs at the base of the dam. However, the scope of study excluded the reservoir body which is crucial because dams are built at the end of the valley, directly above the fault line to prevent river erosion of the valley when there is a tectonic lift due to seismic excitation. In this work, hypothetical seismic time-histories are adopted to replicate potential seismic excitation propagating from ruptured fault lines belonging to the Crocker fault system. ANSYS© ACADEMIC is utilized to simulate the Response Spectrum Analysis (RSA) via finite element method.

2. Modelling of Murum dam

Murum dam in figure 1. (a), is a large concrete gravity dam located at Murum river in Belaga, Sarawak, Malaysia. It is designed to withstand Operating Basis Earthquake (OBE) which is equivalent to 0.1 g PGA shaking or earthquake which has about 50% probability of occurrence during the service life of 100 years, with a return period of 145 years. The design is based on the ICOLD (2010) [11]. However, to understand the performance of RSA, 5% damping is excluded from the simulation. The parameters of the dam are as in figure 1. (b). The height is 141 m, the length along the crest section of the dam is 473 m and the width transverse to the riverbed is 75 m. The reservoir capacity is 12,000 hm^3 and the depth is 135 m. The numerical modelling and meshing with tetrahedral element are constructed using ANSYS© ACADEMIC. The dam and the reservoir are fixed at the bottom as shown in figure 1. (c) and the dam structure is defined to be a target for contact simulation to recreate the reservoir impact on the dam during seismic excitation. In figure 1. (d) the dam structure and the reservoir are at equilibrium. To instruct recognition of ground acceleration input, the definition of pressure direction for every working pressure, P_i at the bottom of the numerical model for is expressed by equation (1):

$$\overrightarrow{P_i} = \frac{(P_jX + P_kY + P_lZ)}{(P_j^2 + P_k^2 + P_l^2)^{\frac{1}{2}}} \quad (1)$$

Where, X, Y, Z are the components of unit vectors normal to centre of element; *PX*, *PY* and *PZ* are the working pressures at X,Y and Z directions, and the values *i,j,k* and *l* represent surface nodes. The dam structure is defined as concrete with Young's Modulus of 31.2 GPa and Poisson's Ratio of 0.18. The density of the reservoir water is 0.998 gcm^{-3}.

3. Static, Modal and Respond Spectrum Analysis

This work employs RSA to visualize the possible seismic damage when Murum dam is loaded beyond 0.1 g PGA [10,12], simulating a magnitude 6.0 earthquake. The initial conditions of the Murum dam is calculated via 3-dimensional static analyses followed by modal analysis. RSA combines the modal analysis with the hypothetical seismic spectrum to generate displacement and stress data. Figure 2 shows the time-histories employed for the RSA which are adopted from Aydin and Tosya earthquakes. Only single-point response spectrum is considered for this simulation.

IOP Conf. Series: Materials Science and Engineering **910** (2020) 012012 doi:10.1088/1757-899X/910/1/012012

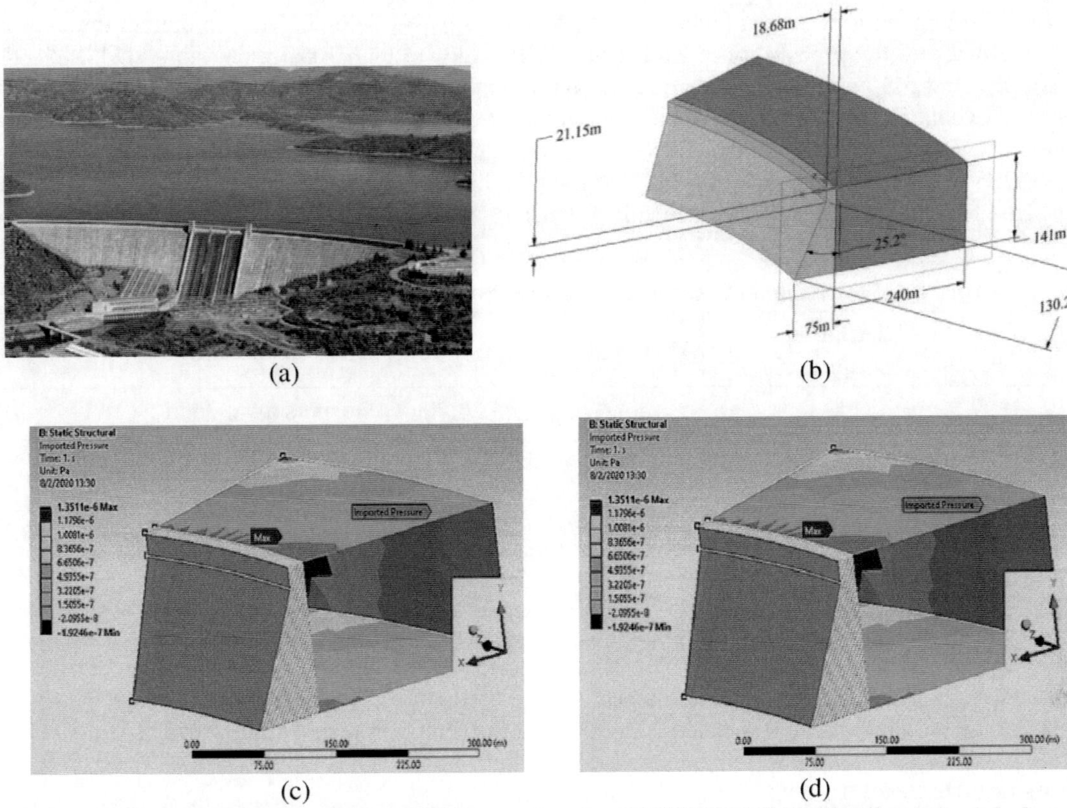

Figure 1. The numerical modelling of Murum dam by ANSYS© ACADEMIC. (a) Aerial photo of Murum dam reservoir at Murum river. (b) The dimension of Murum dam and part of the reservoir. (c) Meshed numerical model with definition of load and fixtures. (d) Cross-section contour of the static reservoir pressure on the dam structure shows that it is at an equilibrium.

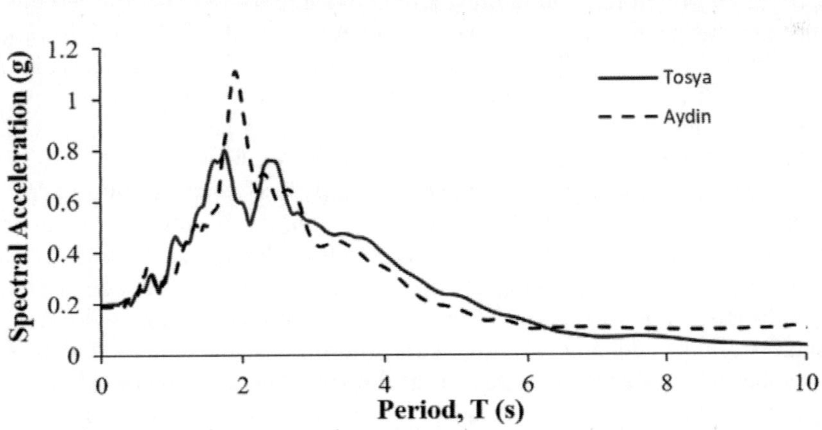

Figure 2. The seismic response spectra of Tosya and Aydin earthquake adopted as time-history input for spectrum analysis.

3.1. Static Analysis

The static analysis is similar to the pseudo-static analysis, whereby the magnitude vector of the load is broken down into components. However, the earthquake load is not defined by the factor value of the dam's dead weight. In this work, the magnitude of the earthquake load is defined as the PGA load itself. Therefore, the discussed results are direct feedback of the seismic excitation. The conditions of the static analysis are featured in figure 1. (c). At the loading node, indicated by the upright arrow, acceleration loadings in g PGA term, where 1g PGA = 9.81 ms^{-2} is defined. Murum dam is in a region of 0.03 g PGA where the probability of exceedance (PE) is lower than 10% in 50 years [13]. In 1994, magnitude 4.7 and 5.3 earthquakes occurred in Kapit – approximately 120 km radius from the dam [14]. These events fit the simulated recurrence frequency within the Crocker Fault system with magnitude ≥5 with return period about 20 years [6]. Therefore, the suitable loadings for RSA are: 0.08 g, 0.1 g, 0.225 g, 0.425 g and 0.625 g PGA. The relationship of PGA loadings in comparison with the Richter Scale Magnitude and Modified Mercalli Intensity scale (MMI) are presented in Table 1.

Table 1. Relationship between PGA, Richter Scale Magnitude and MMI Scale.

PGA (g)	Richter Scale Magnitude	MMI Scale	MMI observations
0.039 - 0.092	4 - 5	V	Sleepers awakened
0.092 - 0.18	5 - 6	VI	Trees sway, some damage from falling objects
0.18 - 0.34	6	VII	Cracking of walls
0.34 - 0.65	6 -7	VIII	Some damage to buildings

3.2. Modal and Respond Spectrum Analysis

Modal simulation is conducted after the static simulation. Then, at the final load step, time-history is defined as input at the loading node. The time-histories from Aydin and Tosya are chosen as hypothetical seismic loadings to replicate the long-wave earthquakes because the maximum PGAs are within the range of magnitude 6.0 which are suitable for the region of responsibility in this study.

4. Results and Discussion

All the extracted results are from the deformation contours at the cross-section of Murum dam. Notable deformation at the lower part of the dam is detected when the load is 0.425 g PGA. Deformation about the value of 0.04 m can be observed at the mid-crest section in figure 3. The RSA results for seismic load at 0.625 g PGA (period 1.5s for Aydin and 1.6s for Tosya) exhibit greater value for the distribution of deformation ranging from 0.04 m to 0.07 m around the mid-crest section. Compared to our previous studies, the deformation yields are smaller [10,12]. It should be reminded that the results are generated from RSA without considering the 5% damping. This shows that the inclusion of reservoir introduces water compressibility that reduces the hydrodynamic pressure and add natural damping to the dam-reservoir-foundation system. Which means the water component absorbs the seismic load during static deformation as explained by Lin et. al. [15]. Overall, values obtained from both static and RSA agree with the crack model for seismic failure of concrete dam [16] and the crack profile of Konya dam [17]. From the results, the tensile zone generates from within the reservoir bed, extending to the heel of the dam foundation. High equivalent stress accumulates at the bottom of the dam because the foundation of the concrete dam is rigid and unable to deform copiously compare to the high deformation of the mid crest section. The deformation and stress values exceed the tensile resistances of concrete structure [18]. This points out there is a likelihood that the mid crest and foundation of the dam are severely damaged at 0.625g PGA as shown in figure 4 and figure 5. These results not only exhibit potential structural failures of Murum dam due to seismic hazard from magnitude ≥5.0 earthquake with the return period about 20 years but also the damage from magnitude ≥6.0 earthquake with the return period 150 years.

5. Conclusion

In conclusion, Murum dam is designed to withstand not more than 0.1g PGA seismic excitation. Therefore, notable hazard patterns are observed when the structure is introduced with hypothetical seismic loads that replicates the plausible earthquake originating from the Crocker fault system. The inclusion of reservoir body into the RSA envisaged clearly that the development of tensile zone is more rapid at the reservoir bed compare to dam foundation itself due to hydrodynamic pressure. The contrast of high deformation at mid crest and high stress at the dam-reservoir bed indicates the likelihood of seismic failure at 0.625g PGA shaking.

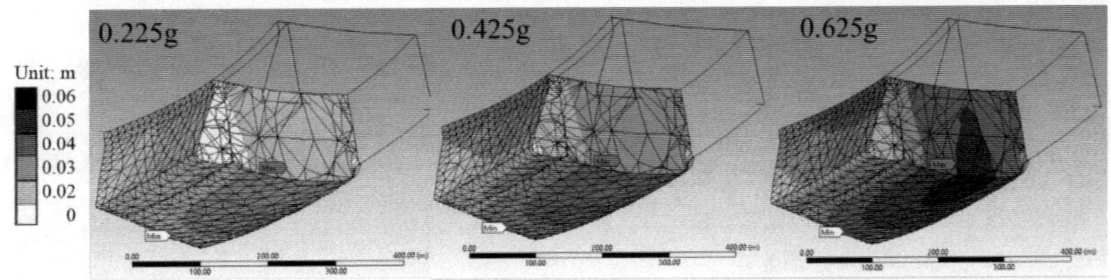

Figure 3. Static deformation of Murum dam at seismic loading 0.225g, 0.425g and 0.625g

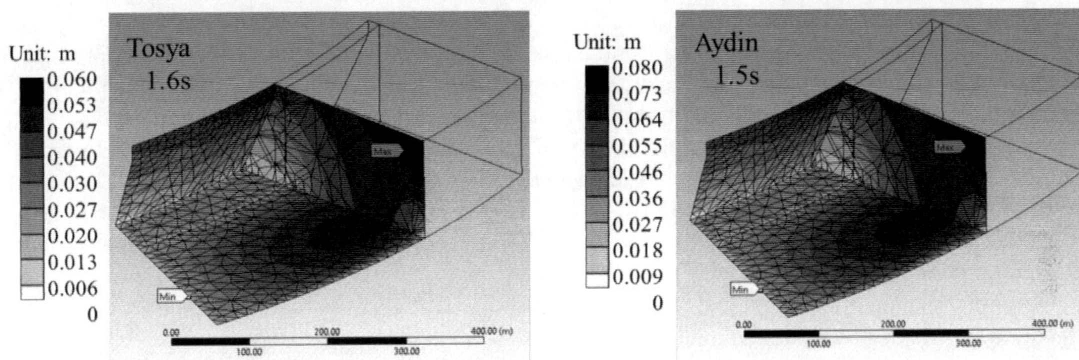

Figure 4. Spectrum analysis deformation of Murum dam at 0.625g for Tosya and Aydin seismic load.

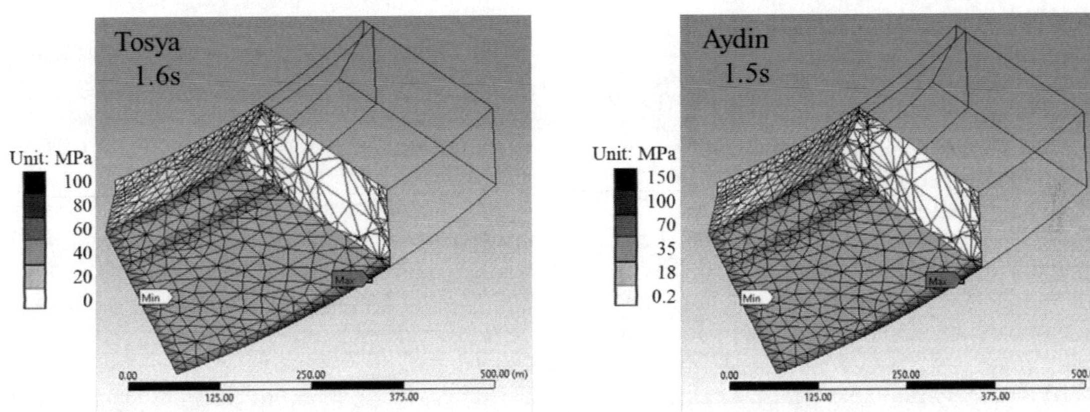

Figure 5. Equivalent stress distribution of Murum dam at 0.625g for Tosya and Aydin seismic load

6. References

[1] Hall R, van Hattum MW, Spakman W. Impact of India–Asia collision on SE Asia: the record in Borneo. *Tectonophysics*. 2008 Apr 28;451(1-4):366-89

[2] Sapin F, Pubellier M, Lahfid A, Janots D, Aubourg C, Ringenbach JC. Onshore record of the subduction of a crustal salient: example of the NW Borneo Wedge. *Terra Nova*. 2011 Aug;23(4):232-40

[3] Shah AA, Zhafri MN, Delson J, Navakanesh B. Major Strike-Slip Faults Identified Using Satellite Data in Central Borneo, SE Asia. *Geosciences*. 2018 May;8(5):156

[4] Malaysian Meteorological Department (MMD) Ministry of Energy, Science, Technology, Environment and Climate Change (MESTECC) 2018 *Asian Disaster Reduction Center (ADRC) Kobe, Japan. COMPARATIVE STUDY ON EARTHQUAKE AND TSUNAMI EARLY WARNING SYSTEM IN MALAYSIA AND JAPAN A Research Study Report for the Visiting Researcher -*

Asian Disaster Reduction Center FY2018 [Internet]. 2018 Nov; Available from: https://www.adrc.asia/aboutus/vrdata/finalreport/FY2018A_MYS_fr.pdf

[5] Chopra AK, Chakrabarti P. The Koyna earthquake and the damage to Koyna dam. *Bulletin of the Seismological Society of America.* 1973 Apr 1;63(2):381-97

[6] Wang Y, Wei S, Wang X, Lindsey EO, Tongkul F, Tapponnier P, Bradley K, Chan CH, Hill EM, Sieh K. The 2015 M w 6.0 Mt. Kinabalu earthquake: an infrequent fault rupture within the Crocker fault system of East Malaysia. Geoscience Letters. 2017 Dec;4(1):6

[7] IEM Position Paper Committee on Earthquake *IEM POSITION DOCUMENT 2005 Approved by IEM Council POSITION PAPER ON ISSUES RELATED TO EARTHQUAKE* [Internet]. 2005. Available from: http://www.myiem.org.my/download/downloadlink.aspx?fn=6852_Position-paper-Earthquake-r2e.pdf&id=6852

[8] Hornblow S, Quigley M, Nicol A, Van Dissen R, Wang N. Paleoseismology of the 2010 Mw 7.1 Darfield (Canterbury) earthquake source, Greendale fault, New Zealand. *Tectonophysics.* 2014 Dec 10; 637:178-90

[9] Chen CH, Saito A, Lin CH, Liu JY, Tsai HF, Tsugawa T, Otsuka Y, Nishioka M, Matsumura M. Long-distance propagation of ionospheric disturbance generated by the 2011 off the Pacific coast of Tohoku Earthquake. *Earth, planets, and space.* 2011 Jul;63(7):881-4

[10] Ahmadi SEISMIC RESPONSE OF MURUM CONCRETE GRAVITY DAM LOCATED IN SARAWAK, MALAYSIA *Conf.: 5th Int. Symp. on Dam Safety at Istanbul, Turkey* 2018

[11] ICOLD SELECTING SEISMIC PARAMETERS FOR LARGE DAMS. Guidelines. *ICOLD Bulletin No. 72 2010* [Internet]. 2010 Available from: https://www.icold-cigb.org/userfiles/files/CIRCULAR/CL1794Annex.pdf

[12] Sim KB Seismic fragility assessment of an existing HEP dam in Sarawak [Master]. Universiti Malaysia Sarawak; 2018

[13] Harith NS, Adnan A, Shoushtari AV. Seismic hazard assessment of east Malaysia region. InInternational Conference on Earthquake Engineering and Seismology (IZIIS-50) 2015 May (pp. 12-16).

[14] Earthquacktrack Recent Earthquakes Near Kapit, Sarawak Malaysia *Earthquacktrack* [Internet] Available from: https://earthquaketrack.com/my-11-kapit/recent

[15] Lin G, Hu ZQ, Du JG, Wang Y, Liu JY. Effects of reservoir boundary absorption on the earthquake response of arch dams. *In Proc. of the 14th World Conference on Earthquake Engineering. Beijing: WCEE* 2008

[16] Hariri-Ardebili MA, Seyed-Kolbadi SM, Mirzabozorg H. A smeared crack model for seismic failure analysis of concrete gravity dams considering fracture energy effects. *Struct Eng Mech.* 2013 Jan 1;48(1):17-39

[17] Azizan NZ, Majid TA, Nazri FM, Maity D, Abdullah J. Incremental Dynamic Analysis of Koyna Dam under Repeated Ground Motions. *In IOP Conference Series: Materials Science and Engineering* 2018 (Vol. 318, No. 1, p. 012021). IOP Publishing.

[18] Kumar P. Earthquake-Resistant Design of RC Structures. In Textbook of Seismic Design 2019 (pp. 179-214). Springer, Singapore

Acknowledgements
This research is adopted and supported by the Disaster Prevention Joint Laboratory under the Japan-ASEAN Science, Technology and Innovation Platform (JASTIP-NET 2019) funded by Kyoto University.

Direct Displacement-Based Design (DDBD) applied to dual wall-frame buildings and steel concentric braced frames

W. Ccahuana[1], J. Coronel[1], and J. Soto[1]
[1] Universidad Peruana de Ciencias Aplicadas

pcciesot@upc.edu.pe

Abstract. One of the most efficient seismic design methodologies is Direct Displacement Based Design (DDBD) because it considers the inelastic range of the structure. There are several investigations on the application of this methodology in frames, structural wall and bridges, but those related to steel structures with concentric bracing are scarcer. The objective of this investigation is to determine the damping parameters and the seismic responses of a dual concrete building and one of concentrically braced steel frames, to later compare their base loads and displacements. The procedure first determines the design offset for the 1GDL model; then the desired responses are obtained by this displacement. This methodology uses an effective stiffness, which is composed of an elastic and an inelastic state of the structure to determine seismic responses.

1. Introduction

The large number of earthquakes in the Ring of Fire and its high population density forces Civil Engineering to investigate and propose new and more precise methods when making the seismic design of a structure.

A first reference is that of Priestley et al. [1], which established the theoretical guidelines on Displacement Based Design (DBD) with a special methodology called Direct Displacement Based Design (DDBD), which is what will be used in this work.

Subsequently, research was carried out on the application of this methodology in steel frame buildings. Connections and concentric bracing (CBF) [2] in the structures have a great impact on the damping of these, which is demonstrated by the article. A structure with CBF type bracing has a great increase in lateral stiffness, so its displacement profile is much smaller than one without bracing [3]. All values were corroborated with the American Society of Civil Engineers [4].

With respect to the application of the DDBD method in buildings with double wall frames, the influence of the cutting walls on the displacement profile of the structure is used. In addition, unlike steel frames, in a dual wall-frame building with cracks there is an increase in its damping due to the cracked inertia of its structural elements.

In this research, two buildings of similar architectural distribution have been analyzed using this seismic design methodology, but with different structural systems: dual wall-frame buildings and steel concentric braced frames. Since this design only applies to axes and not to the structure as a whole, the most influential and best represented axis for the design in both structural systems has been chosen. The damping in both cases will be obtained from the guidelines proposed by Priestley [1]; however, some modifications have been made, corresponding to investigations that support a more specific damping. Finally, the answers are presented as shear and basal moment which will be compared and analyzed.

2. Method

Seismic analysis will be carried out using the DDBD methodology to two structural systems following the guidelines established by Priestley [1], to then compare the results in displacements and basal forces. The structure's damping behavior will be included through the results of previous investigations [5], [6], [7]; Damping is an important parameter, since it is the one that most influences the seismic response.

The damping used in structural systems are the following:

- Dual wall-frame system:

$$\zeta_W = 0.05 + 0.444\left(\frac{\mu-1}{\mu\pi}\right) \tag{1}$$

Source: Priestley Et al. 2007

- Steel frames system:

$$\zeta_W = 0.05 + 0.565\left(\frac{\mu-1}{\mu\pi}\right) \tag{2}$$

Source: Priestley Et al. 2007

Where the parameter μ represents the ductility of system displacement.

3. Results

To validate the investigation, two structural systems were used, which are: Dual system (Figure 1) and steel frame system (Figure 2).

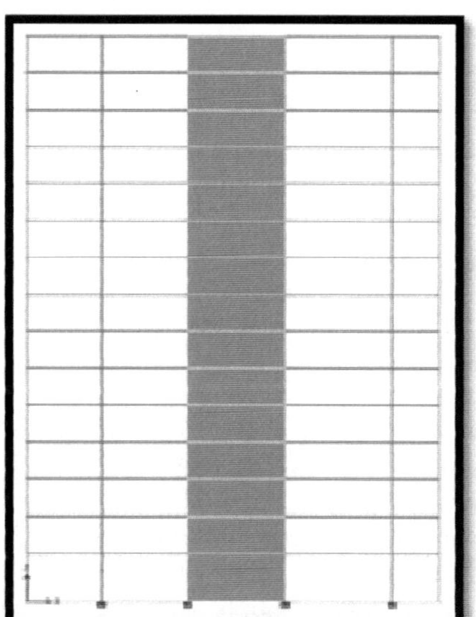

Figure 1. Dual Structure (Source: Own elaboration in Etabs)

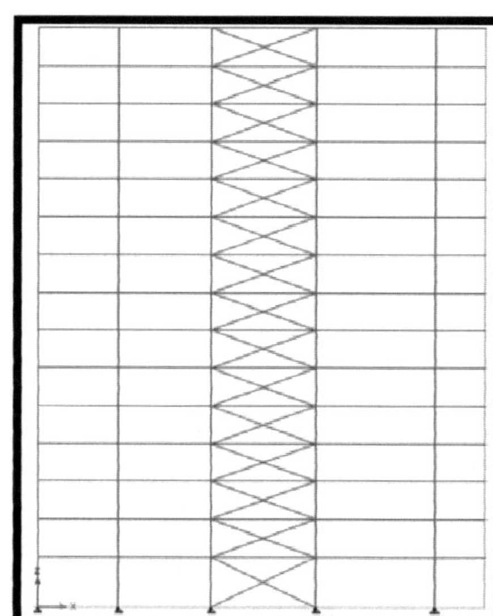

Figure 2. Steel frame structure (Source: Own elaboration in Etabs)

The comparison will be between 2 buildings of 15 levels with the same architectural configuration. The first level will be 4 meters high and the following ones will be 3 meters high. The live loads in each system will be 450kgf / m2 for a tributary width of 7m in the beams and an independent dead load per level specified in the following paragraphs.

The dual structural system has a dead load of 600 kgf/m2 on the first level and 500kgf/m2 on the following levels. The structural elements will have the following sections: 60x80 cm columns, 30x50 cm beams and 30cm cutting walls. The amount of steel in the structural elements will be evaluated through the Etabs program. The concrete used is 280kgf/cm2 in verticals and 210kgf/cm2 in horizontals. The amount of steel used will be 1% for columns and beams 0.0018.

The structural system of steel frames has a dead load of 400kgf / m2 on the first level and 300kgf/m2 on the following levels. The structural elements will have the following sections: HSS20x20 metal columns, T 15x15 beams and HSS8x8 bracing. The connections will be of the cut type.

3.1. DDBD seismic response in dual system

The theoretical guidelines followed in this research are those proposed by Priestley [1]. Next, the results obtained under the DDBD method and non-linear Pushover analysis method are presented in order to compare them. A drift limit state of 2.5% damage prevention is considered.

The dual structural system that uses the DDBD methodology has a baseline shear of 1772kN, greater than the non-linear methodology by 54.26% as seen in Figure 3. A linear tendency is observed in Figure 4. The story moments at the base are 53927kN-m, greater than the non-linear system by 50.98%, the latter being 36536 kN-m present in Figure 5. The maximum drift obtained with the DDBD method is 2.5%, controlled by the limit state of damage control and a maximum drift of the non-linear methodology at 0.44% as seen in Figure 6.

The displacement through the DDBD is 110 cm in the limit state of damage prevention compared to the 4.4 cm obtained from the non-linear methodology observed in the capacity curve of Figure 7. These results are due to the simplicity of the DDBD methodology that does not consider the hysteretic effect of the structure and, consequently, greater displacements are acquired.

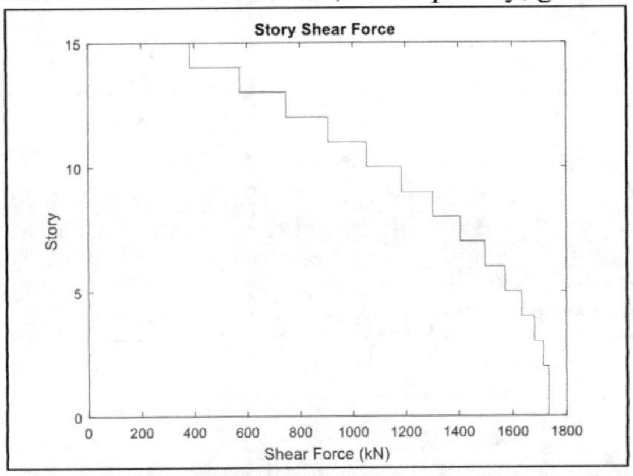

Figure 3. DDBD shear forces (Source: own source)

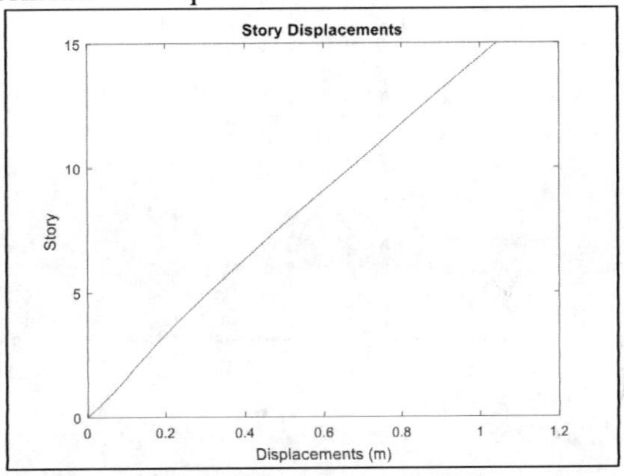

Figure 4. DDBD displacements (Source: own source)

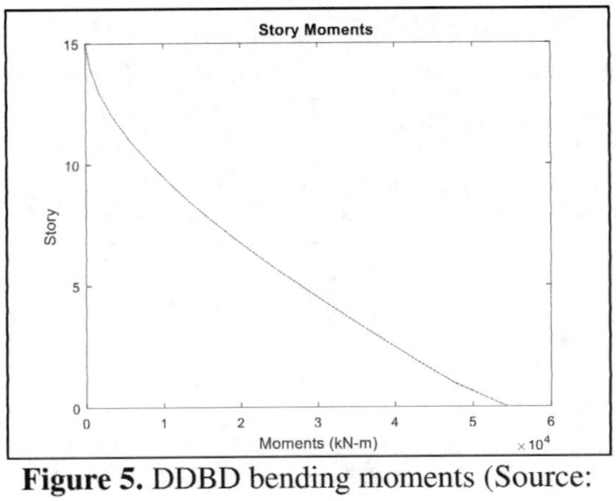

Figure 5. DDBD bending moments (Source: own source)

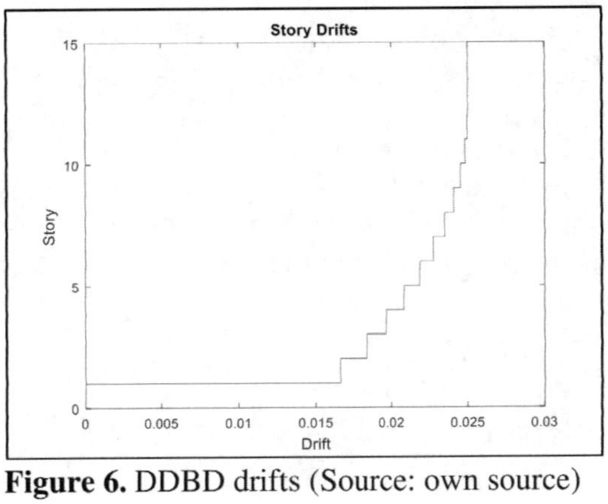

Figure 6. DDBD drifts (Source: own source)

It should be mentioned that for the non-linear Pushover methodology, hinges were used at 0.1 and 0.9 of the length of the columns and beams obtaining an effective stiffness of 2829 kN/m. In contrast, in the DDBD methodology an effective stiffness of 2475 kN/m was calculated, this being inferior to that obtained in the non-linear analysis. This imprecision implies a considerable increase in the forces and, consequently, in the sections of the structural elements.

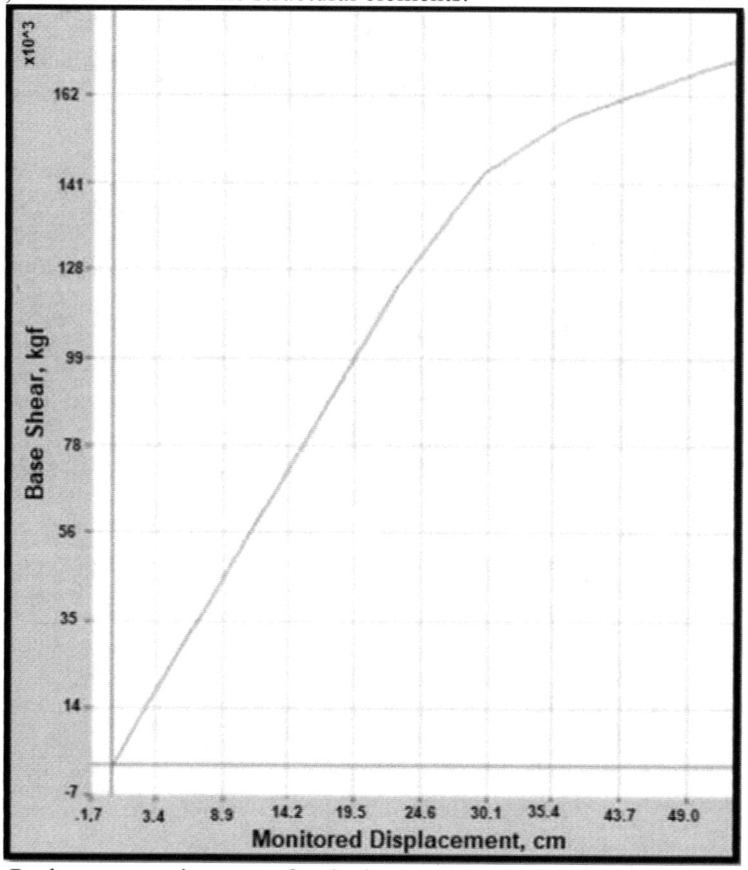

Figure 7. Pushover capacity curve for dual system (Source: own elaboration in Etabs)

3.2. DDBD seismic response in steel frames system

In the same way as in the dual system, the theoretical guidelines followed in this investigation are those proposed by Priestley [1]. Next, the results obtained under the DDBD method and non-linear Pushover analysis method are presented in order to compare them. A drift limit state of 2.5% damage prevention is considered.

The steel frame system using the DDBD methodology presents a baseline shear of 2024 kN, which is greater than the non-linear methodology by 54.26% and is observed in Figure 8. The story displacements are shown in Figure 9.

The moments at the base are 18927 kN-m, greater than the non-linear system by 50.98%, the latter being 12536kN-m present in Figure 10. The maximum drift obtained with the DDBD method is 2.5% controlled by the damage control limit state and a maximum drift of the non-linear methodology at 0.44% as it is shown in Figure 11.

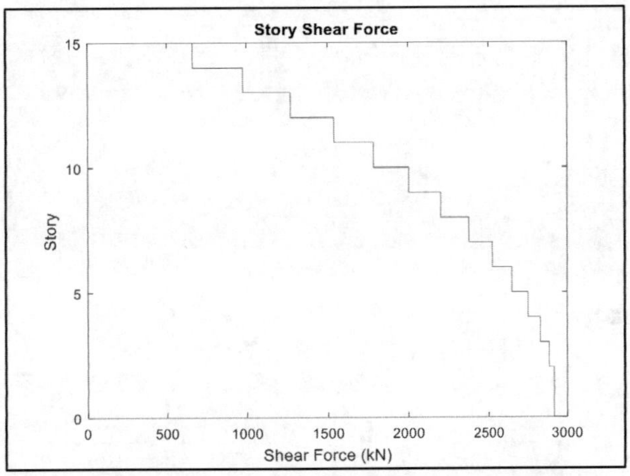

Figure 8. DDBD shear forces (Source: own source)

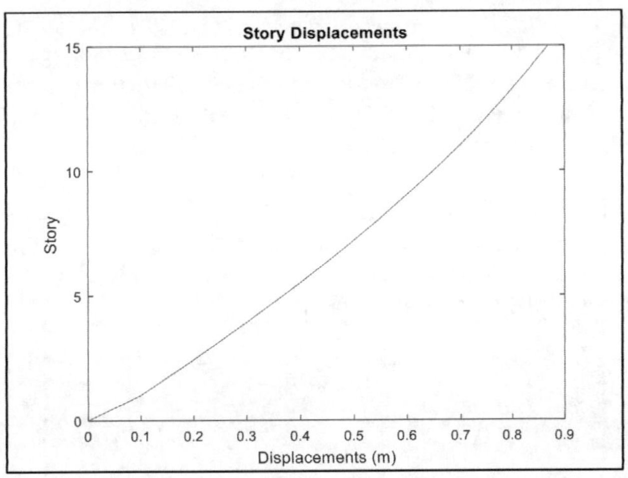

Figure 9. DDBD displacements (Source: own source)

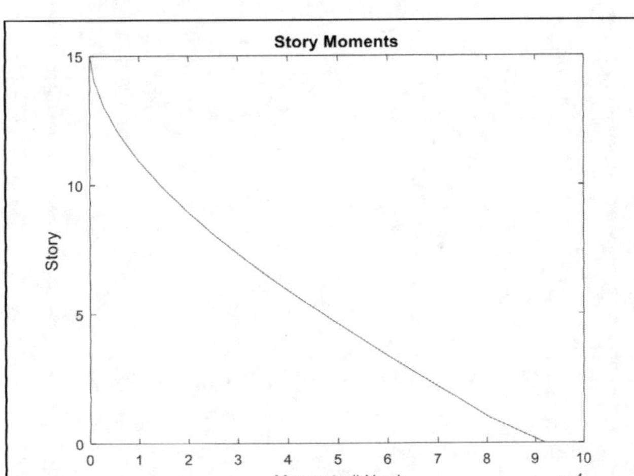

Figure 10. DDBD bending moments (Source: own source)

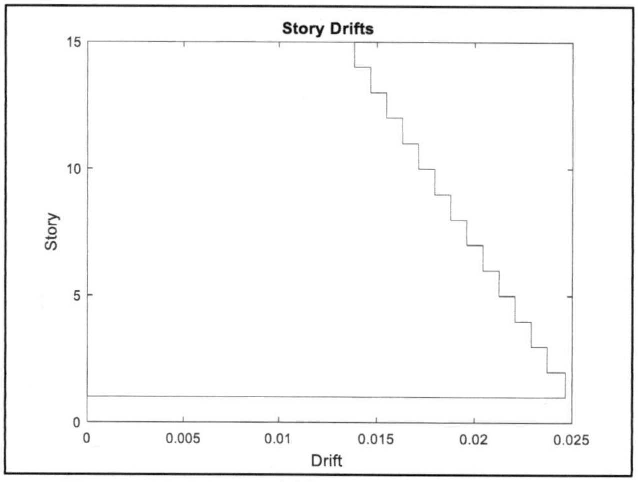

Figure 11. DDBD drifts (Source: own source)

The displacement using the DDBD is 118 cm in the limit state of damage prevention compared to the 4.4cm obtained from the non-linear methodology observed in the capacity curve of Figure 12. These results are due to the simplicity of the DDBD methodology that does not consider the hysteretic effect of the structure and, consequently, greater displacements are obtained. It is necessary to mention that for the non-linear Pushover methodology, hinges were used at 0.1 and 0.9 of the length of the columns and beams obtaining an effective stiffness of 4532 kN / m in the capacity curve. Instead, in the DDBD methodology calculated an effective stiffness of 3248 kN / m, this being inferior to that obtained in the non-linear analysis. This imprecision implies a considerable increase in the forces and, consequently, in the sections of the structural elements.

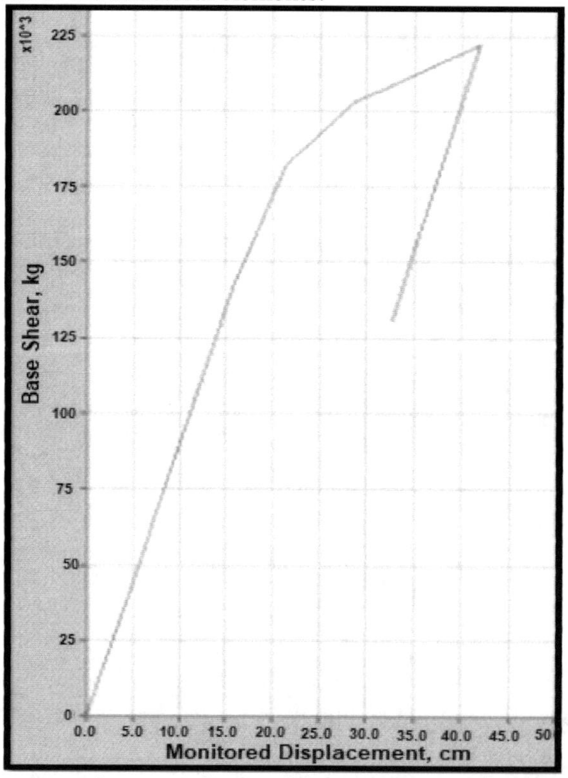

Figure 12. Pushover capacity curve for Steel frames (Source: own elaboration in Etabs)

4. Conclusions

The displacements are the same in the two structural systems by condition of the DDBD methodology, in which it is established that the maximum displacement in the two structures will be 2.5% of the height of the building per damage prevention limit state according to Priestley[1].

The basal shear force in the steel frame is greater than the dual system in 6%. In ideal conditions, being the most ductile steel structure, it allows greater displacements. However, the system is not ideal and has a damping that decreases its effective period and consequently its basal forces. The period obtained in the contributed steel system was 3.96 s and 5.48 s in the dual system.

The damping in the steel structure is 13% and that of the dual system is 18%. The damping in a dual system increases due to the cracking of the structural elements of the concrete and in a system provided with steel it increases mainly due to the buckling of its steel columns.

The DDBD methodology allows to avoid oversizing in structural design of the structure. This is due to the fact that this methodology falls within the inelastic range of the structure and we do not assume that all its behavior is linear, which confirms its proximity to the real effects with the comparison with the capacity curve obtained from the non-linear Pushover analysis. When it is considered that the behavior of a structure is always linear, it is when oversize begins to exist due to an excessive increase in seismic forces. This happens because non-linear behavior, such as cracking the structure, releases efforts and makes the ultimate design efforts less.

5. References

[1] M. Priestley, G. Calvi and M. Kowalski, "*Displacement-Based Seismic Design of Structures*" IUSS PRESS.

[2] S. Salawadeh, T. Ryan, B. Broderick and J. Goggins, "*DDBD assessment of steel CBFs using full scale shake table tests with realistic connections*" Journals of Construction.

[3] M. Ferraioli and A. Lavino, "*A Displacement-Based Design Method for Seismic Retrofit of RC Buildings Using Dissipative Braces*" Mathematical Problems in Engineering

[4] *Seismic Evaluation and Retrofit of Existing Buildings* (ASCE 41-13).

[5] R. Abadi and O. Bahar, "*Investigation of the LS Level Hysteretic Damping Capacity of Steel MR Frames Needs for the Direct Displacement-Based Design Method*" Korean Society of Civil Engineers, pp 1-12.

[6] D. Ranjan, A. Prakash, "*Seismic behaviour of concentrically braced frames designed using direct displacement-based method*" International Journal of Steel Structures.

[7] M. Al-Mashaykhi, P. Rajeev, K. Wijesundara, M. Hashemi, "*Displacement profile for displacement based seismic design of concentric braced frames*" Journal of Constructional Steel Research.

Acknowledgments

This research has been possible due to constant support from the Civil Engineering Faculty at the Engineering Department of the Universidad Peruana de Ciencias Aplicadas (UPC) in Lima, Peru.

IOP Conf. Series: Materials Science and Engineering **910** (2020) 012014 doi:10.1088/1757-899X/910/1/012014

Digital photogrammetric survey from aerial images for architectural documentation in sites and monuments: a case study of Wat Mahathat Lop Bury, Thailand

M. L. Varodom Suksawasdi[1,2]

[1] Faculty of Architecture and Planning, Thammasat University, Pathumthani, Thailand.
[2] Research Cluster in Livable Environment and Architectural Design Laboratory (LEAD-Lab)

Email: varodom@ap.tu.ac.th

Abstract. This study is a part of research to describe a way to obtain various photogrammetric products from aerial photograph using a drone. The aim of the project was to develop a methodology to obtain information for a case study of the architecture documentation in Wat Mahathat Lop Bury, Lop Bury Province, Thailand which a combining the manoeuvrability and low cost of a drone with the accuracy of the results of the Agisoft PhotoScan software. These will be improvements for architecture documentation working with limited budgets or with limited access to sites.

1. Introduction

The architecture documentation must be have accuracy and alignments of each structure which to analyze data such as accurate measurements of dimensions, structure, construction and technology. The architecture documentation use to preservation and restoration for revitalization sites and monuments [1]. Maps of Thailand sites and monuments available in the literature have an insufficient precision to perform such research; consequently, for the study of each architectural site, a precise survey must be done [2]. Several solutions are available for architectural survey such as a topographical survey can produce accurate measurements that can be processed with software to obtain maps and 3D models. Nevertheless, with a total station, it is impossible to register each stone and deformation of the ruins. Thus, the representation is a synthesis and an interpretation of the information. A second solution would be using a laser scanner. Unfortunately, archaeological sites can be quite widespread and have many monuments, so the number of stations to be performed and the number of point cloud to be processed would demand too much post processing resources. A third solution is to use a drone to acquire aerial photographs of sites, and process them to create orthophotos and 3D point clouds.

2. Developments of Photogrammetric product in Architecture documentation

Photogrammetric product has existed for over a hundred years [3]. But just expanding to architecture two decades ago. Knowledge resulting from the combination of scientific principles and the arts in relation to measurement technology and photography. In the field of architecture conservation is applied to surveying buildings with limitations such as access, primitiveness and risk of damage is being withdrawn. The two important steps are photography and creating orthophotos as shown in Figure 1. First step have a shooting methods, image preparation and image distortion correction before processing and Second step have a process uses principles to establish relationships of objects, directions, and distances from images, which can be referenced using geometric theories. The results obtained are processed. And advanced calculations Nowadays, there are complex computer software which their will get three-dimensional images, coordinates, object locations, sizes, shapes, proportions and distances of buildings or locations [4].

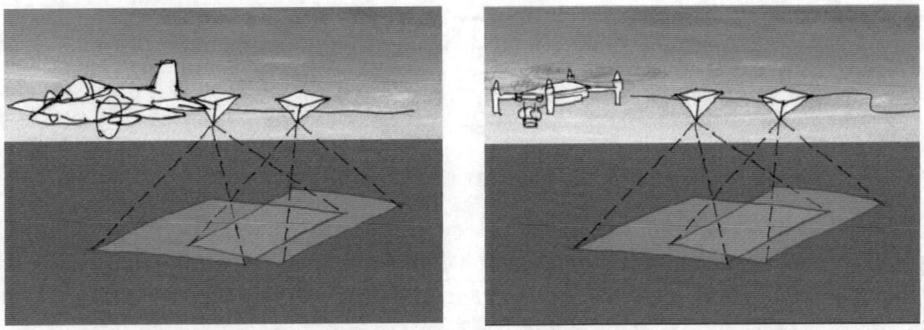

Figure 1. Adjacent but overlapping aerial photos are called stereo-pairs and are needed to determine parallax and stereo/3D viewing.

Cultural heritage is a complex and highly political phenomenon. There are few social elements and types of tourism that are more hotly contested at so many levels. These groups have at some point in history been oppressed by the ethnic group in power, which has resulted in their pasts having been deemphasized or even written out of official history in some cases while many countries are embarrassed by the cruelties performed against racial minorities (and sometimes majorities) throughout history, special interest groups and proponents of heritage commemoration are forcing legislators and other public officials and organization's to acknowledge the atrocities of the past and commemorate them [5]. Cultural heritage tourism limitations that it had been prevented tourists from participating in areas or activities. Mostly it is caused by personal factors and external factors. These could be divided into 4 limitations; accessibility, education preparation, disability and psychological constraints [6]. These limitations will be reduced by presenting 3D models which are created by the process of photogrammetry.

3. Research methodology

Study and collect data from flying with Dji Mavic 2 pro drone with a sensor 1" CMOS, FOV Lens about 77°, maximum flight time (no wind condition) about 31 minutes and find suitable procedures for flight. By flying around the object to collect images in each of the images circling and assessments to use of this system in comparison between the use to 3D Scaner. The desired result is the point cloud formats for both systems. Finally, look for limitations in studying this process and for further study in the future shown in Figure 2.

The review also none highlighted the need for any operation to be fully compliant with aviation regulations and health and safety provisions, to be covered by comprehensive insurance and aware of privacy issues (See Table 1). We also consulted external organisations with more experience in using UAVs in sites and monuments recording, and the consistent advice was that it would be risky to purchase an expensive 3D scanner without first experimenting using Professional.

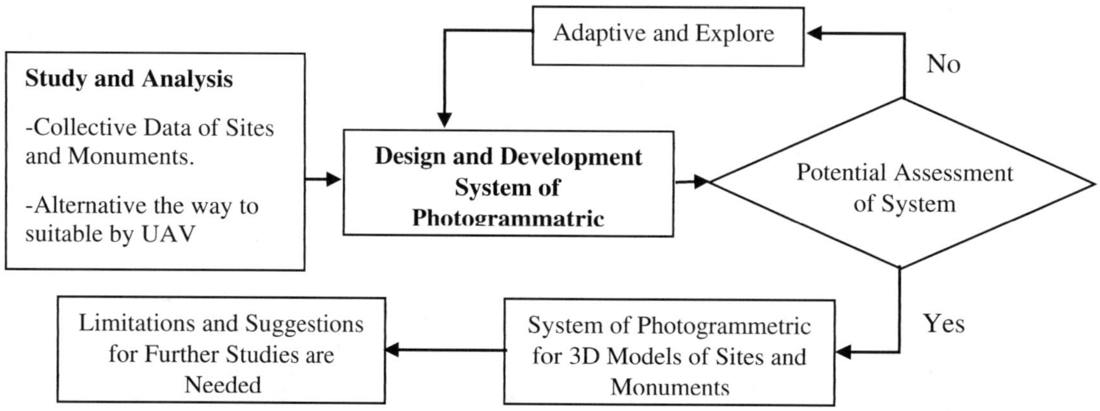

Figure 2. A framework of this study.

Table 1. Options review for Historic Environment Thailand sites and monuments unmanned aerial vehicles (UAV) photographic applications [2].

Options	Pros	Cons
Buy UAV	In Thailand, readily available at short notice	Cost of equipment, maintenance, training, developing competence, operations manual
	Flexibility to use at short notice	Must be used regularly to maintain operator competence
	Sites and Monuments imperatives drive image capture	Potential damage/total loss of equipment
	Takes less time	Insurance and public liability
	Suitable for both Sites and Monuments	Rapidly developing field, legal framework likely to change
Buy 3D scanner	High quality imagery	More expensive than UAV
	High accuracy (Color of material)	Requires staff time (ideally 3 operators)
		Not Suitable to small work.

4. Results

The sites and monuments considered is situated at Mueang Lop Buri of Lop Buri province shown in Figure 3. It was the religious center of a Lop Buri ancient city and had been reduced to the role of the central location. These resulted in a flying height of 30 meters from the ground (in Thailand restricted fly more than 90 meters from the ground). The maximum rate of the photographs was chosen (Around to object) and a flying speed of 5 m/s was calculated. The exposure time used was 1/320 sec, the ISO 100, aperture 5 and the focal length 10.26 mm. The flight plan consists of around to object. The flight editor announces a flight time of 20 min. between flights the UAV lands in order to change the battery sets.

To obtain orthophotos and 3D models, Agisoft PhotoScan software, was used. I got 26 images that were processed to obtain the aerial photograph shown in Figure 4. The software starts searching the reference point between the images, before calculating the distortion parameters and thus the calibration of the camera. It turned out that the 3D point cloud representing the tie points presents a bending which is not in accordance with the reality [4]. After this the commands for the creation of orthophotos and 3D Models can be launched.

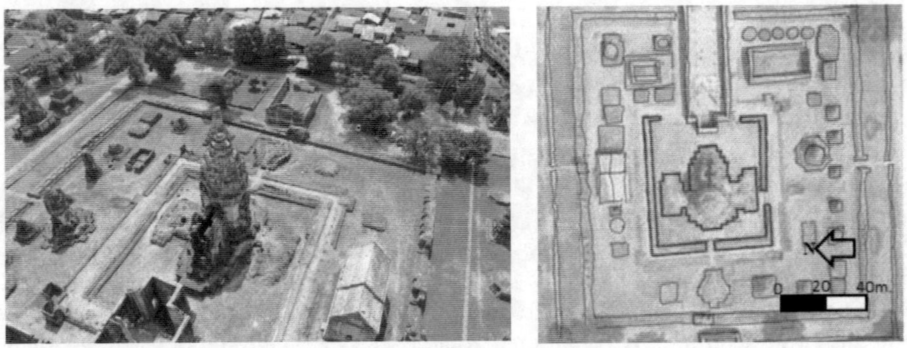

Figure 3. Show photos at Wat Mahathat Lop Buri are taken on 2 March 2019.

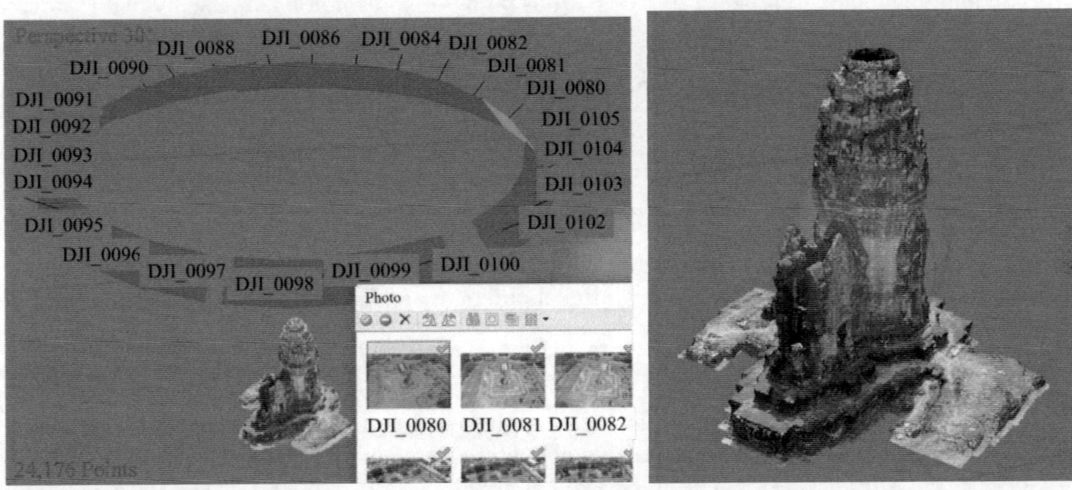

Figure 4. Show photos in time shot mode using the DJI Mavic2 drone and continue processing it in an Agisoft PhotoScan software.

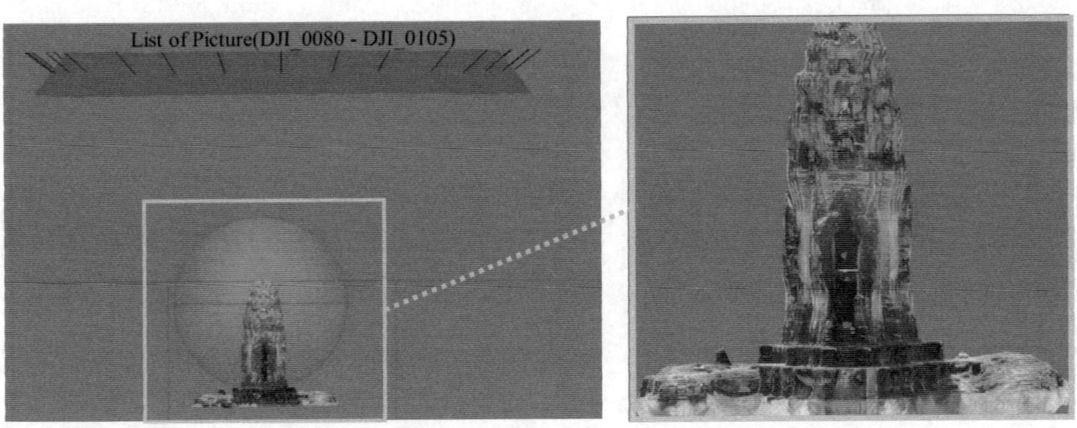

Figure 5. Show the introduction of results that can tell the correct proportion (true shape) of archaeological sites.

5. Conclusions

This study can applied in presentations to display interesting exhibits in the museum and stimulates the mind to become increasingly concerned about the disappearance of the national heritage. Such as Understanding the concept of the cosmology and architectural planning from this concept [7]. For future generations to see imaging reconstruction and cultural landscape of architecture that has been lost. In the presentation with 3D modeling, if the building is complete, architectural documentation by the survey makes too easy to know the true distance shown in Figure 5. Therefore, the case study is archaeological sites and buildings are incomplete. And the materials that make up the structure are bricks and laterite. Photogrammetry Easier to do. On the other hand, images are presented with 3D Modeling. When brought into an application to display architectural images. It can open the side view (elevation) and cut view (section). By doing this, be able to understand the physical characteristics of that archaeological site. For example Building proportion Decorative elements or missing parts. It also speculates the characteristics of the structure. The style of construction and materials, the original color of the building. The discrepancy of calculations with Agisoft PhotoScan Also available to create 3D Modeling as in the Table 2.

Table 2. Geometrical accuracy error calculated by Agisoft PhotoScan [8].

Count	X error (cm)	Y error (cm)	Z error (cm)	XY error (cm)	Total (cm)
32	0.47	0.51	0.68	0.69	0.98

As for the geometrical accuracy, it has already been found in several projects that photogrammetry, in combination with using targets, is able to product very accurate result [8]. Further of this study had been helping decision between participant and stakeholders in architectural conservation and improved planning for building preservation.

6. References

[1] Charnnarong S 2012 Photogrammetry in architectural conservation: Literature review and possible applications for Thai traditional architecture *NAJUA* 9 pp 159-184

[2] Cowley DC, Moriarty C, Geddes G, Brown GL, Wade T and Nichol CJ 2018 UAVs in Context: Archaeological Airborne Recording in a National Body of Survey and Record *Drones* 2(1) 2

[3] Burtch R 2006 *History of Photogrammetry* (The center for Photogrammetric Training) Ferris State University

[4] Mouget A and Lucet G 2014 Photogrammetric Archaeological Survey with UAV *ISPRS Annals of Photogrammetry, Remote Sensing and Spatial Information Sciences* 2(5) 258

[5] Timothy DJ and Boyd SW 2006 Heritage tourism in the 21st century: Valued traditions and new perspectives *Journal of heritage tourism* 1(1) pp1-16

[6] Timothy DJ and Boyd SW 2003 *Heritage tourism* (Harlow England) New York Pearson Education

[7] Jirathasanakul S 2001 *Wat: Buddha of Thai Architecture* (Thammasat Printing House) Bangkok

[8] Waagen J 2019 New technology and archaeological practice. Improving the primary archaeological recording process in excavation by means of UAS photogrammetry *Journal of Archaeological Science* 101 pp11-20

Acknowledgement

The authors gratefully acknowledge the financial support provided by Thammasat University Research Fund under the TU research Scholar, Contract No. 125/2561 and this research has been approved by a research ethics committee of Thammasat University.

IOP Conf. Series: Materials Science and Engineering **910** (2020) 012015 doi:10.1088/1757-899X/910/1/012015

Modeling of Conceptual Public Building Design Factors to enhance Safety of Tourists in Thailand

Porntip Ruengtam

Faculty of Architecture, Urban Design and Creative Arts, Mahasarakham University 44150 Thailand, porntip.r@msu.ac.th (+66) 805034298

porntip.r@msu.ac.th

Abstract. A conceptual public building design framework was modeled using architectural and universal design concepts to enhance the safety of tourists in Thailand. The research method involved informal interviews and observations for the area case study of Koh Yo, Songkhla, Thailand. Interview results were summarized together with related theories from the literature. All factors were collected and listed. Ten experts in related fields confirmed the factors and framework through the technique of a focus group discussion. Findings indicated that tourist characteristics had a positive direct effect on requirements for the safety of public buildings and comprised three sub-factors including (1) general architectural facilities, (2) warning systems, and (3) fire safety and building regulations. Each sub-factor affected tourist safety satisfaction. This framework can be used to quantitatively analyze future policy for the design and management of public buildings to enhance the safety of tourists in Thailand.

1. Introduction

Tourism is one of the main industries that brings foreign exchange income to Thailand. The country is a popular tourist destination with a large diversity of tourism resources. According to the National Statistics Office [1], the tourism industry has a gross domestic product (GDP) of around 6.00%. The tourists mostly hail from China followed by Malaysia, Korea, Laos, Japan, India, the United Kingdom, Australia, the United States and Russia, respectively. Safety and security are important issues for the development of tourism destinations. Tourists are naturally concerned about their safety and security when traveling, and often have scant knowledge of their intended destination compared to their home country [2]. Tourists are aware that they may be at risk of physical injury due to various hazards when traveling including potential insecurity such as crime, terrorism or accidents [3]. The issue of tourist safety in Thailand must be addressed to promote tourism capability and support sustainable future growth [4]. Ko Yo is an island surrounded by Songkhla Lake that has beautiful scenery, old temples, unique local foods and many tourist attractions. The island is suitable for promoting tourism by the community. Most of the people who live on Ko Yo are engaged in fishing and trading processed fishery products. Ko Yo also has a thriving woven fabric industry producing well-known products. A conceptual research model of tourist characteristics and their requirements for safety in public buildings was formulated using the case study area as Koh Yo, Songkhla, Thailand. To achieve this objective, components of tourist requirements were first identified and then measured. A conceptual model was developed to demonstrate the relationships of factors between tourist characteristics, their safety requirements, and their attitudes toward safety. The model was proved by collecting empirical data for quantitative analysis. Results can be used to assess the quality of design and management of public buildings to enhance the safety of tourists in Thailand.

2. Literature review

Architectural design involves creating buildings and facilities as cost-effective to meets the needs of users [5]. Architectural design must also take into account the safety of the building users such as accidents from falling on a slippery floor, walls collapsing and sufficient fire exits [5]. Safety can be defined as the protection of people against unintended consequences of an involuntary nature. For example, a case of arson is a security issue, while a spontaneous fire is a safety issue [2]. Safety in public buildings is the key issue in this study. A public building is used by many people each day. Therefore, the safety and stability of the building is the most important matter, regardless of the type or purpose of use. Facilities are tangible items as part of a property or building, especially as added extras that make people more comfortable. Facilities are basic requirements for tourists such as accommodation, buildings, parks, swimming pools, exercise rooms, signboards, communication systems, public utilities, transportation and other services. Guidelines for preventing and managing safety for foreign tourists require strong cooperation of stakeholders from all sectors including government, private sector, local communities and tourists [6]. Guidelines for tourism safety management can be stated as follows: (1) Enforcing the law and imposing penalties for serious offenders because good law enforcement will encourage tourism [7]. (2) Development of safety at tourist destinations by supporting tools and equipment to help tourists including CCTV, machine translation or signage translated into many languages, light installation in tourist attractions and basic tourist equipment [8]. (3) Promotion of hygiene in tourist destinations by improving public toilets, waste management and wastewater treatment [4]. Fire safety and building regulations require the availability of necessary protective equipment in the event of an accident. Equipment such as fire extinguishers, heat and smoke detectors and emergency lighting must pass inspection standards and be ready for use [9]. Universal design requires that any product or environment must involve the consideration of many factors including aesthetics, engineering options, environmental issues, safety concerns, industry standards, and cost [10]. The National Disability Authority [11] has defined universal design as the design and composition of an environment so that it can be accessed, understood and used to the greatest extent possible by all people, regardless of their age, size, ability or disability. The universal design concept was considered and combined in the analysis of this study. The Center for Universal Design [12] has established seven principles of universal design to provide guidance in the design of products and environments. These principles include.

- Equitable use - the design is useful and marketable to those with diverse abilities.
- Flexibility in use - the design accommodates a wide range of individual preferences and abilities.
- Simplicity and intuition - use of the design is easy to understand, regardless of the user's experience, knowledge, language skills, or current concentration level.
- Perceptible information - the design communicates necessary information effectively to the user, regardless of ambient conditions or the user's sensory abilities.
- Tolerance for error - the design minimizes hazards and the adverse consequences of accidental or unintended actions.
- Low physical effort - the design can be used efficiently, comfortably, and with a minimum of fatigue.
- Size and space for approach and use - appropriate size and space is provided for approach, reach, manipulation, and use, regardless of body size, posture, or mobility.

3. Research methods

The technique of focus group discussion [13] is a widely used and accepted method for achieving convergence of opinion concerning real-world knowledge solicited from a group interview within certain topic areas. The technique is designed as an expert group interview process and aims to conduct detailed examinations and discussions of a specific issue. The process is continuously iterated until a consensus is determined to have been achieved. This research used the focus group discussion technique to confirm the factors involved in the conceptual modeling of public building design to enhance the safety of tourists in Thailand. Steps of the focus group discussion technique were as follows:

Step 1. Identify the factors: Observations and informal interviews were conducted with tourists in the case study area (Koh Yo, Songkhla, Thailand) in terms of general management and architectural designs of safety in public buildings. Interview results were summarized along with related theories from the literature such as textbooks, research articles (ISI, SCOPUS and ScienceDirect databases) and related annual reports in Thailand. All factors were collated and listed before proceeding to the next step.

Step 2. Confirm the factors: The listed factors were confirmed by ten experts [13] in related fields using the focus group discussion technique. The experts were selected based on their professional experience as follows: Five experts from tourists in the case study, three experts in architectural design aspects of public buildings employed by private design companies, and two experts from local tourism government officers in Koh Yo, Songkhla, Thailand. The listed factors were confirmed by a consensus of opinions starting from expert 1, 2, 3..., 10 and returning to expert 1 again. All experts carefully considered the listed factors. Results were aggregated and shared with the group after each round until a consensus was saturated. Once the listed factors were confirmed and categorized by the ten experts, they were constructed as the conceptual model in the next section.

Step 3. Draft the conceptual model: According to the expert opinions, confirmed factors were categorized and arrows were drawn between the factors as a draft conceptual model. This was then reconfirmed by the ten experts by repeating Step 2. The final model was considered as the conceptual research model.

4. Results

From the focus group discussion, the opinions of the ten experts as confirmed factors and items as indicators of the study are listed in Table 1.

Table 1 shows the factors of tourist characteristics comprising age and health as TC1 and TC2. Tourist requirements comprised three sub-factors as (1) general architectural facilities; 10 items (Q1 to Q10), (2) warning systems; 4 items (Q11 to Q14), and (3) fire safety and building regulations; 9 items (Q15 to Q23), while tourist safety satisfaction comprised 3 items (TS1 to TS3). All factors for tourist requirements and their measurement items were used to construct the conceptual model shown as Figure 1. This model was extended to include more detail of the items, variables and their indicators as shown in Figure 2.

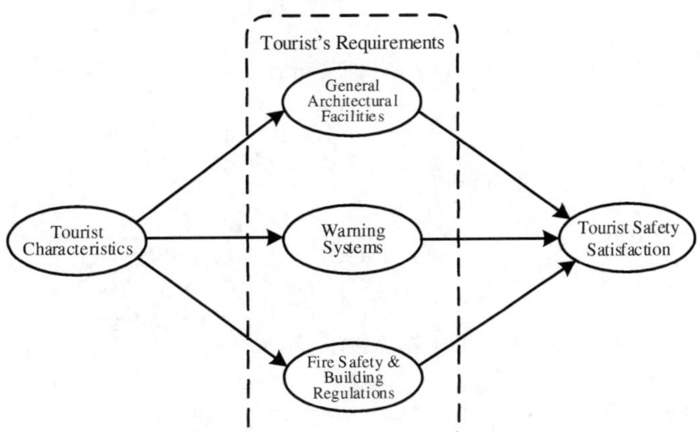

Figure 1. Conceptual model for public building design factors to enhance the safety of tourists in Thailand.

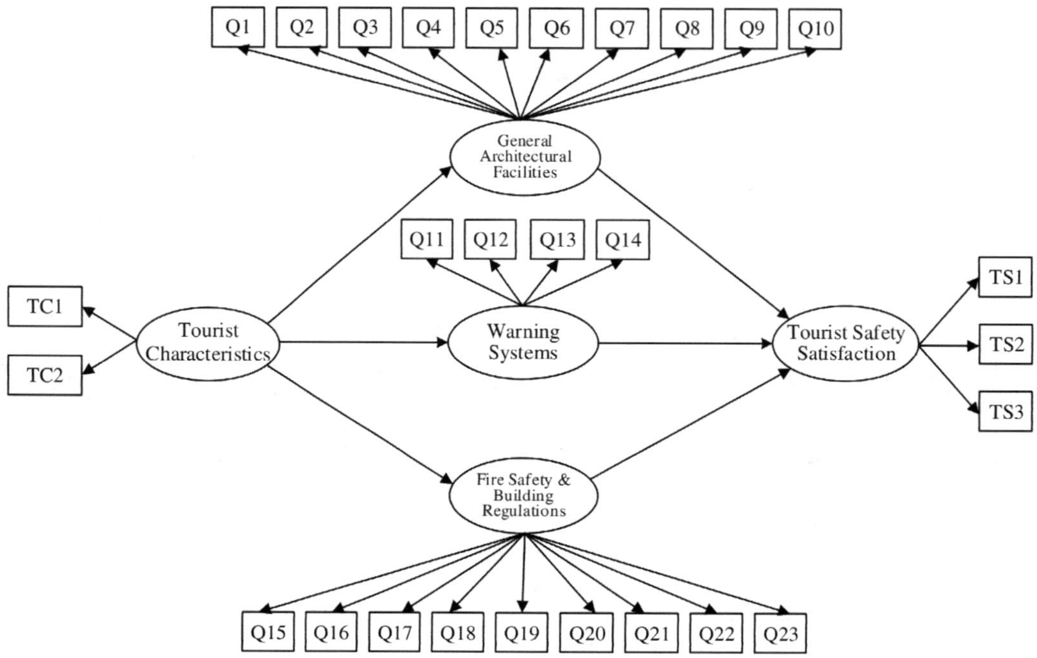

Figure 2. Detailed conceptual model

Table 1. Factors for tourist's requirements and their measurement items

Factor		Item
Tourist characteristics		TC1 Age TC2 Health
Tourist requirements	General architectural facilities	Q1 Floor/different level floor Q2 Stair Q3 Ramp Q4 Fall protection wall / hand rail Q5 Door / window Q6 Elevator Q7 Escalator Q8 Lighting Q9 Car park Q10 Toilet
	Warning systems	Q11 Warning sign board / symbol Q12 Notification sound Q13 CCTV Q14 Safety officer
	Fire safety and building regulations	Q15 Fire extinguisher Q16 Heat and smoke detector Q17 Emergency lighting Q18 Fire sprinkler Q19 Fire alarm Q20 Fire cabinet Q21 Fire exit Q22 Assembly area Q23 Local regulation
Tourist safety satisfaction		TS1 Confident in the safety TS2 Reliable in the safety TS3 Satisfying in the safety

5. Conclusions

This research studied factors related to enhancing the safety of tourists in public buildings by modeling a conceptual framework. The case study area (Koh Yo, Songkhla, Thailand) was observed, and interviews were conducted to assess the general management and architectural design for safety in public buildings. Factors were identified from the case study and confirmed by ten experts using the focus group discussion technique. Results indicated that requirements of tourist characteristics included age and health as TC1 and TC2, general architectural facilities Q1-Q10, warning systems Q11-Q14, fire safety and building regulations Q15-Q23, and tourist safety satisfaction TS1-TS3 as shown in Table 1. A conceptual model as the outcome from the study (Figure 2) showed that factors of tourist characteristics had a positive direct effect on their safety requirements. These requirements comprised three sub-factors including (1) general architectural facilities, (2) warning systems, and (3) fire safety and building regulations. All factors impacted on tourist safety satisfaction in public buildings in the case study area. The model in Figure 1 was extended to include more details of the items, variables and indicators as shown in Figure 2. This model can be used to quantitatively analyze future policy for the design and management of public buildings to enhance the safety of tourists in Thailand.

6. References

[1] Burgstahler S 2015 Universal Design: Process, Principles, and Applications (Washington: University of Washington, College of Engineering, UW Information Technology, College of Education)

[2] Center for Universal Design 2021 Universal design principles [1 Mar 2021] Available from https://www.ncsu.edu/ncsu/design/cud/about_ud/udprinciples.htm

[3] Committee for Design, Construction and Inspection 2011 Design standards for buildings and locations (Bangkok: Office of Physical Systems Management Chulalongkorn University)

[4] Emmitt S, Prins M and Otter A D 2009 Architectural Management International Research and Practice (Chennai: Newgen Imaging Systems Pvt Ltd)

[5] George R. 2003 Tourist's perceptions of safety and security while visiting Cape Town. Tourism Management, 24(5), pp.575-585

[6] Grudens-Schuck, N, Allen B L and Larson K 2004 Methodology Brief: Focus Group Fundamentals 12 (Iowa: Extension Community and Economic Development Publications)

[7] Jeff W and Stephen J P 2003 Managing Tourist Health and Safety in the New Millennium: Current Status of Tourist Health and Safety 1st ed. (Oxford: Elsevier Science Ltd.)

[8] Mansfeld Y and Pizam, A 2006 Tourism, security and safety: from theory to practice (Oxford: Elsevier Butterworth-Heinemann)

[9] Ministry of Tourism and Sports. 2017 Tourism Development Plan 2 (2017-2021) (Bangkok: The War Veterans Organization of Thailand press)

[10] National Disability Authority 2021 10 things to know about UD [1 Mar 2021] Available from Information on http://universaldesign.ie/What-is-Universal-Design/The-10-things-to-know-about-UD

[11] National Statistical Office 2019 Statistical Yearbook Thailand 2019 [1 Jan. 2021] Available from http://service.nso.go.th/nso/nsopublish/pubs/e-book/SYB-2562/files/assets/basic-html/index.html#1

[12] Tarlow P E 2014 Tourism security: strategies for effectively managing travel risk and safety (Waltham: Elsevier Inc.)

[13] Tarlow P E and Santana G 2002 Providing safety for tourists: a study of a selected sample of tourist destinations in the United States and Brazil. Journal of Travel Research, 40(4), pp. 424-431

Acknowledgments

This research project was financially supported by Mahasarakham University, Thailand.

IOP Conf. Series: Materials Science and Engineering **910** (2020) 012016 doi:10.1088/1757-899X/910/1/012016

Analysis in Integrated Design Potentials Achieving Nearly Zero Energy in Office Buildings in Bangkok Neighborhood

K Lohwanitchai [1] and D Jareemit[1]

[1] Faculty of Architecture and Planning, Thammasat University, Pathumthani, Thailand 12121

Corresponding email: non_ran@hotmail.com

Abstract. This study aims to investigate the energy-saving potentials and design approaches to move the small and medium office buildings in Bangkok neighborhood areas to become the Net-Zero Energy Building (NZEB). The envelope designs, such as increasing the material's thermal resistance, adding shading devices, and using high energy-efficient systems, together with the electricity production from the photovoltaic cell, are applied in the studied buildings. According to the energy simulation results, with the implementation of current design techniques, the annual energy consumption of office buildings cannot reach the NZEB. However, these design approaches can enforce the maximum building's energy-saving capability at 88 kWh/m^2 that is nearly the Economic Building (ECON). The maximum energy reduction occurs when improving the high thermal performance of the building's envelope and using high energy-efficient systems. To achieve this challenging target. Future studies should account for more energy-efficient systems as well as the building's operation schedule.

1. Introduction

In Thailand, the office building, as included in the commercial sector, consumes electricity consumption by 35 percent of Thailand's energy consumption, after the industrial sector [1]. The rapid growth of office building from the past decades causes that energy demand has been increasing continuously. This increment of energy demand is a major contributing factor to global warming. Furthermore, a limit of natural energy sources could affect the energy security of the country in the future. With this concern, the Ministry of Energy has set the Thailand's 20-year Energy Efficiency Development Plan (EEDP) that all the new government and private buildings, including office buildings, have to be more energy efficiency and become the NZEB by 2030 [1]. However, the design guidelines, as well as energy-saving approaches for architects and engineers to design the office buildings achieving the NZEB, is still limited.

There is a possible approach to improve the office buildings to become NZEB for hot and humid conditions. According to the reviews of previous projects [2-6], the design for NZEBs comprises two main parts: 1) decreasing building energy use, and 2) generating electricity from renewable energy resources. The decrease in building energy consumption can be done in several ways. Most studies started with passive design techniques, including building design and material selection. Those techniques are involved in architectural work. The second technique is a part of the engineering task that the higher efficient performance of mechanical ventilation and lighting systems are selected. The final approach is to manage the building operation schedule, which requires a good corporation from the building owner, building manager, and occupants. This study aims to investigate the design potential that introduces the office building to become the zero energy in hot and humid conditions. Several approaches recommended in the previous works are applied in the building case studies to minimize the annual energy consumption. Then the potential electricity generation from renewable energy systems is implemented. The study's results can provide a position of energy consumption performance of the office buildings and a design guideline for improving the buildings being higher energy efficiency.

2. Methodology and case study buildings

Figure 1 presents the framework of this research. This study selects two different building areas of 5,000 m^2 (31x31m) and 15,000 m^2 (50x50m), which are mostly found in typical office buildings, as case study buildings. The office buildings are located in the Bangkok neighborhood areas, which both buildings are a six-story building. The building height is less than 23m due to the local regulation. The floor to floor of each building has a 3-meter height.

Based on design approaches to be the NZEB recommended in the previous works, in this study, decreasing the annual energy consumption focuses on two main approaches: 1) building's envelope design and 2) energy-efficient system design. First is to reduce the amount of heat transfer through the building envelope by increasing the thermal resistance of opaque walls (scenario A) and windows (scenario B). Then, providing shade a building to prevent the envelope directly exposed to the solar heat (scenario C). After the design approaches are applied, in scenario D, the high efficient ventilation system with a higher Energy Efficiency Ratio (EER) value and variable airflow volume (VAV) is assigned. The LED light bulb is used to minimize energy consumption from the lighting demand. It is appointed that the air conditioner of the studied buildings is operated from 8 a.m. to 5 p.m. in the weekday, while there is no operation during the weekend.

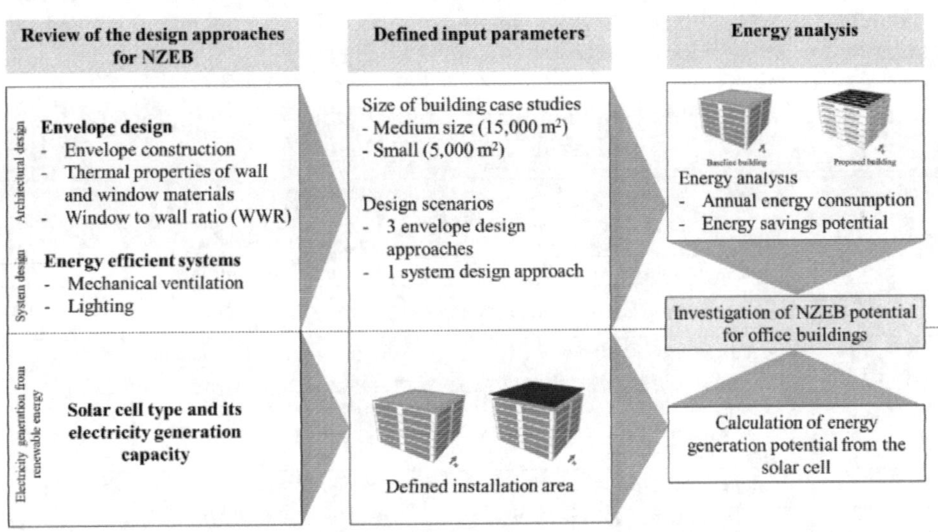

Figure 1. Research framework

Table 1 presents the setting conditions of the baseline and proposed buildings, which include thermal properties of wall and window materials, window to wall ratio (WWR), shading device's length, air-conditioning system, and operation schedule of air conditioning units. The annual energy use in office buildings is performed via using eQuest model [7] under Bangkok weather data [8]. As mentioned above, the office buildings are in Bangkok neighborhood areas. As a consequence, the simulated buildings are not affected by surrounded tree shade and adjacent building shade.

Table 1. Design characteristics of the building envelope and air-conditioning system

Material	Baseline building	Proposed building
Wall construction (opaque envelope)	Metal Frame with Fiberglass Insulation (U=0.53 W/m^{2}K)	Metal Frame with Fiberglass Insulation (U=0.48 W/m^{2}K) + Polyisocyanurate for interior isolation (U=0.14 W/m^{2}K)
Wall Construction (internal zone)	Wall type frame (U=2.28 W/m^{2}K)	Wall type frame with fiberglass insulation (U=0.77 W/m^{2}K)
Roof construction	Metal Frame with 3-inch Polyurethane Insulation (U = 0.056 W/m^{2}K)	Metal Frame with 3-inch Polyurethane Insulation (R-18, U = 0.056) + Fiberglass for Interior Insulation (U = 0.048 W/m^{2}K)
Window	Green Laminated (U=5.52 W/m^{2}K, VLT :0.71, SC 0.6)	Insulated Glass (U=1.55 W/m^{2}K, VLT :0.62, SC 0.31)
Window to wall ratio (WWR)	80	20
Shading (Length)	0 m	1.5 m
Air-conditioning system (A/C)	**Baseline building**	**Proposed building**
Cooling Source	DX Coils	DX Coils
A/C System Type	Package Split Type	Package Split Type
EER	12.5	13.76
Supply Fan	1.00 in WG (High)	1.00 in WG (Premium)
Lighting system	**Baseline building**	**Proposed building**
Energy per Area	13.34 W/m^2	5 W/m^2, LED and Task Light

3. Results and Discussion

In the long-term, the EEDP has targeted improving the office buildings in Thailand toward a higher standard of energy-saving capability when compared to the current minimum energy consumption efficiency standard of the buildings (BEC), as shown in Table 2 [1]. To achieve the NZEB, the annual energy consumption in office building needs to below 57 kWh/m^2.

Table 2. Total energy consumption required for an office building under different levels of energy-saving capability [1]

Type	Description	Energy-saving capability (kWh/m^2-year)
Reference	The average value of overall annual energy consumption in buildings	219
BEC (Building Energy Code)	Minimum energy consumption efficiency standard of the buildings	171
HEPS (High Energy Performance Standard)	The high energy efficiency standard of the various system which can be achievable by using current technologies	141
ECON (Economic Building)	The target in the near future when the technologies of equipment and various systems are developed to be more energy efficient but are still cost-effective	82
ZEB (Zero Energy Building)	The long-term target when the need for external energy supply to the buildings is near zero because the energy demand of such buildings is very low and there is also on-site energy generation from renewable energy	57

Figure 2 presents the energy consumptions calculated from each design scenario as compared with that of the reference building and four energy consumption defined for higher energy performance buildings. Overall, the calculated annual energy consumption of the baseline buildings ranges from 88–153 kWh/m^2, which is lower than the reference as defined in the EEDP [1]. The building's energy consumption is mostly from cooling energy use, which accounts for approximately 60 percent. The average annual energy consumption of the small office buildings is 9 percent higher than that of the medium office buildings. The annual energy consumption of the buildings with lower WWR is 11–16 percent lower than that of higher WWR. The primary energy reduction occurs when increasing the wall's thermal resistance (scenario A) for the lower WWR and using a high-performance window (scenario B) for the higher one. Another significant reduction occurs when using high-performance systems (scenario

D). In the final renovation scenario, the annual energy consumptions in the small and medium office buildings show 88–100 kWh/m^2, which are nearly ECON building.

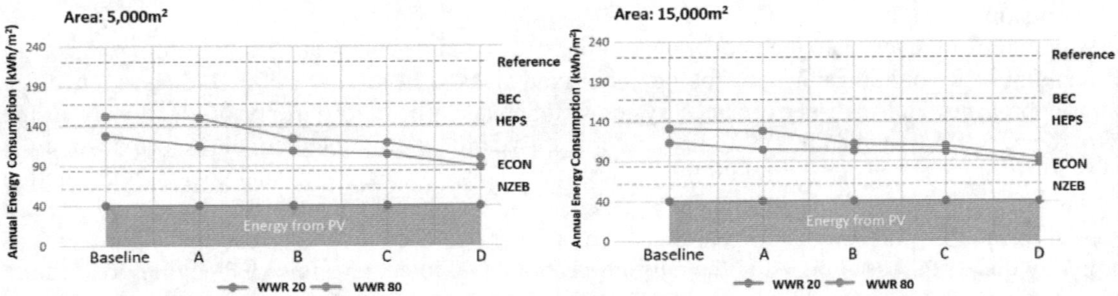

Note: Scenario A is an increase in the wall's thermal resistance. Scenario B is increasing the window's thermal resistance. Scenario C is adding a shading device and Scenario D is improving the higher performance of A/C and lighting systems.

Figure 2. Calculations of annual energy consumption in small and medium office buildings when implied with different renovation scenarios

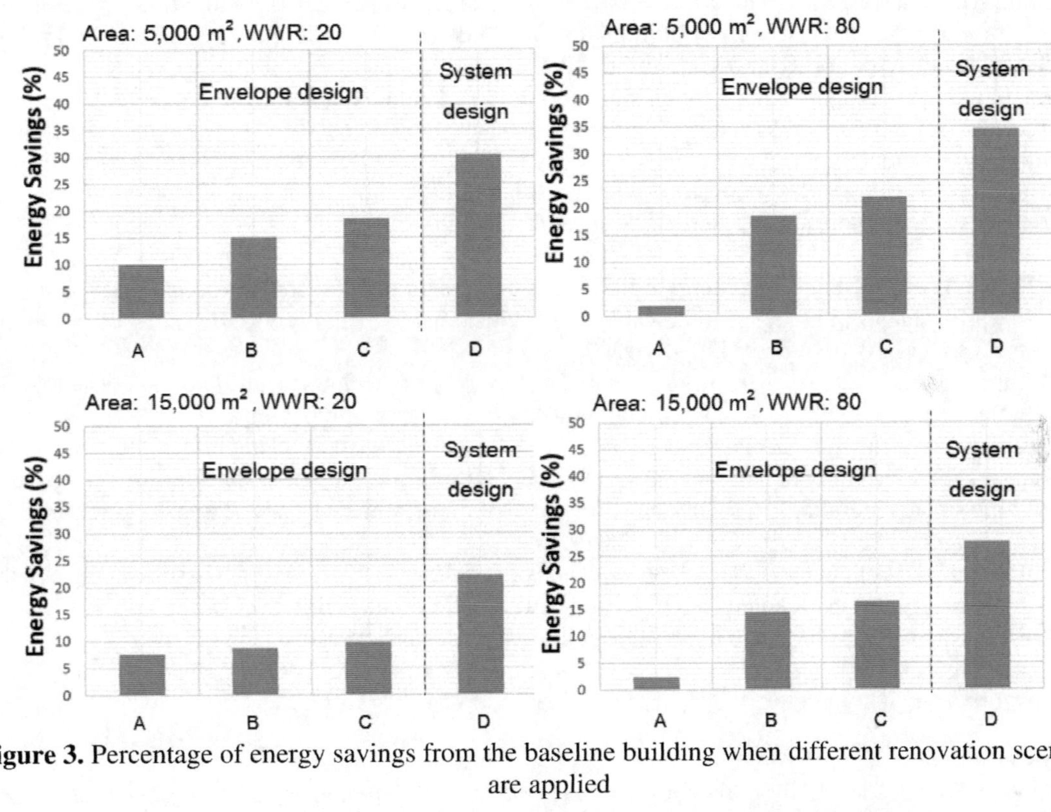

Figure 3. Percentage of energy savings from the baseline building when different renovation scenarios are applied

When considering energy-saving performance, the small office building takes more benefits from the envelope design and the system renovations than that of the medium building do. The maximum savings obtained from the final renovation scenario (scenario D) in the small building accounts for 34 percent. In comparison, the energy savings in the medium office building shows up to 27 percent (shown in figure 3). From those saving potential, the energy reduction of 18–22 percent is achieved by improving the thermal properties of the building envelope. Using high energy efficient window could considerably reduce the annual energy consumption in the buildings with WWR of 80 by up to 16 percent. At the same time, this technique provides small savings in the buildings with lower WWR. In contrast, significant energy reduction from increasing thermal resistance of the opaque wall occurs in cases with lower WWR.

4. Conclusion
This study aims to investigate the energy-saving potentials and design strategies to improve the small and medium office buildings in Bangkok neighborhood areas to become the NZEB. Several envelope design approaches and high-performance systems, together with the electricity production from the photovoltaic cell, are applied in the studied buildings. Even improving the thermal performance of the building envelope and replacing it with high energy-efficient systems, the annual energy consumption of those office buildings are still higher than that recommended for the NZEB. However, there is potential to improve the small and medium office buildings to become higher energy performance as defined by the HEPS based on using the current technologies. In achieve this challenging goal, future studies should account for higher thermal performance windows and energy-efficient systems. Furthermore, building operation schedules should be considered in the design.

5. References
[1] Energy Policy & Planning Office. (2019). Thailand 20-year energy efficiency development plan (2018–2038). Ministry of Energy. Bangkok, Thailand.
[2] Shin, M., Baltazar, J.-C., Haberl, J. S., Frazier, E., & Lynn, B. (2019). Evaluation of the energy performance of a net zero energy building in a hot and humid climate. Energy and Buildings, 204, 109531. https://doi.org/10.1016/j.enbuild.2019.109531
[3] Lasternas, B. (2018). School of Design and Environment National University of Singapore. IGEM-IUSGBC 2018 Toward Net Zero Energy Buildings. http://www.mgbc.org.my/Downloads/20181018_IUSGBC_2018_Towards_Net_Zero_Energy_Buildings/6_Tallest_Net_Zero_Building_in_Singapore_Pushing_the_Limit_of_Net_Zero_in_Tropical_Climate_Bertrand_Lasternas_NUS.pdf
[4] Feng, W., Zhang, Q., Ji, H., Wang, R., Zhou, N., Ye, Q., Hao, B., Li, Y., Luo, D., & Lau, S. S. Y. (2019). A review of net zero energy buildings in hot and humid climates: Experience learned from 34 case study buildings. Renewable and Sustainable Energy Reviews, 114, 109303. https://doi.org/10.1016/j.rser.2019.109303
[5] Yimprayoon, C. (2016). Zero Energy Building. Journal of Architectural/Planning Research and Studies (JARS), 13(2), 1–30. https://so02.tci-thaijo.org/index.php/jars/article/view/81264
[6] Belussi, L., Barozzi, B., Bellazzi, A., Danza, L., Devitofrancesco, A., Fanciulli, C., Ghellere, M., Guazzi, G., Meroni, I., Salamone, F., Scamoni, F., & Scrosati, C. (2019). A review of performance of zero energy buildings and energy efficiency solutions. Journal of Building Engineering, 25, 100772. https://doi.org/10.1016/j.jobe.2019.100772
[7] Hirsch, J. (2010). *EQuest Introductory Tutorial, version 3.64*. James J. Hirsch & Associates. http://doe2.com/download/equest/eQ-v3-64_Introductory-Tutorial.pdf
[8] NREL. (2016). Weather Data by Location. EnergyPlus. Retrieved from https://energyplus.net/weather-location/asia_wmo_region_2/THA//THA_Bangkok .484560_IWEC.

Acknowledgment
We are grateful for special support from Pongpeera Saekow, a manager from Jardine Engineering Corporation, for providing information and recommendations for mechanical ventilation system design.

IOP Conf. Series: Materials Science and Engineering **910** (2020) 012017 doi:10.1088/1757-899X/910/1/012017

A Readiness of Areas and Space Planning in Houses Supporting Emergency Services for Fall Injuries in The Elderly

B Supudomkul[1], D Jareemit[1], P Khanchaitit[2], and T Janjamlha[1]

[1]Faculty of Architecture and Planning Thammasat University, Pathumthani 12121, Thailand
[2]National Nanotechnology Center (NANOTEC), 111 Thailand Science Park, Paholyothin Road, Klong 1, Klong Luang, Pathumthani 12120, Thailand.

Corresponding e-mail: boonsinee.aptu@gmail.com

Abstract. The global population structure is currently stepping into an aging society. In the change of the world population structure, critical care of older adults from fall injuries at home is a major public health issue since the falls can lead to disability or death. By reviewing the architectural design research, most studies focus on the design approaches aiming to reduce the risk of fall accidents occurred among the elderly. However, no study has investigated the correlation of architectural design in supporting the medical care of seniors after slip and fall accidents. Therefore, this research aims to evaluate area and space planning of typical three single houses supporting emergency medical care for falling accidents amongst the elderly. The study firstly investigates the activities and the area required for the medical operation by interviewing medical specialists and rescue teams from two hospitals and two related agencies. Then, the activities and area requirements are mapped with the room layout of three typical single houses located in the Bangkok neighborhood areas. It is found that the designed layout and room area of all studied houses inadequately supports for emergency situations, which has only two available areas from a total of thirteen areas. The details of the space availability and obstacles for emergency medical care are discussed and proposed for improving the space planning and layout of the house for future work.

1. Introduction

The continued growth of the older population increases the possibility of the elderly living alone in their homes and risk from fall accidents [1]. Falls in the elderly are a significant cause of accidental or unintentional injury deaths. Several organizations related to public health, especially the World Health Organization (WHO), Centre of Disease Control and Prevention (CDC) and National Council of Aging (NCOA), have investigated variable risk factors for the falls and proposed effective prevention strategies to prevent and reduce the risk of older adult's falls. Furthermore, many studies have comprehensively studied fall prevention by educating on self-care skills [2], arranging the home environment and providing necessary equipment [3]. At present, Artificial Intelligence (AI) and robots have been developed for elderly care [4-5]. According to the reviews, most studies focus on the strategies preventing falls in older adults. However, there is a research gap on the oversight of the procedure after the fall accidents occur.

Emergency medical service is essential and significant to saves elders from falling and injuring themselves in their homes [6]. The process requires rapid access to treat the patients and ensure comfortable operational space for medical care. In consideration to house plans, the design mainly focuses on its intended function or purpose. However, the operational space for emergency medical care on-site is possibly exclude in the initial design. When fall accidents occur, inadequate layout planning could delay access to healthcare services in houses. This study aims to investigate area requirements for on-site medical care for older adult falls together with the evaluation of the room layout of typical single houses in Thailand that can support the emergency medical services (EMS) to help older adults from falls. The study's findings could provide barriers and limitations of designed space for the operation of medical treatment in Thai houses. Therefore, recommended design and further investigations are proposed.

2. Methodology

The study calculates area requirement and space for rescue staff, response kits, lifting, and the handling of equipment to help older adults who have fallen in their homes. These required rescue area and space is then mapped in the first-floor plan of the studied houses to assess the available space for on-site medical care. The procedure of moving the elderly to an ambulance is not accounted for in this study.

Medical specialists and rescue teams from four organizations were interviewed, including professionals working in Siriraj Hospital, Chulabhorn Hospital, the Ruamkatanyu rescue team, and the First Ambulance team. The questions for the interview include the following: How many staff work for on-site services?; What rescue kits and equipment, including its dimension, are used during on-site medical care?; What activities they do during the treatment of older adults who have fallen? These activities and suggested required equipment from the interview are later used to define the dimension and area of on-site medical care. According to the interview, the devices used for safe and proper relocation of patients into the ambulance comprises of stretcher cots, long spinal boards, stair chairs, and scoop stretchers. During the treatment, the medical team practically utilizes a long spinal board as a first aid and evacuation tool while the emergency bag and accessories are integral for medical treatment. In this study, the area required for the on-site emergency service is calculated based on the performance of medical treatment with the long spinal board since this device requires a larger service area than other devices. The length of the board is 1.85 m in length and 0.45 m in width. Figure 1 presents the area requirement and dimension for on-site medical care. It is assumed that three medical staff is carrying the patient on both sides of the long spinal board. As a consequence, the on-site medical care with a long spinal board requires a minimum area of 7.06 m^2 with a width of 2.05 m, and 3.45 m in length.

Previous study [7] investigated and classified design characteristics and building floor plans of Thai single-detached houses located in neighborhood areas in Bangkok. This study selects three models of single-detached houses, which accounts for a large portion of Thailand's housing stock. It is defined that the studied houses have a bedroom for older adults living on the first floor. The areas that the falls are likely to occur in the house comprises of the bedroom (1), toilet (2), stair (3), corridor 1 (4), corridor 2 (5), living area (6), dining area 1 (7), dining area 2 (8), pantry (9), kitchen (10), washing area (11), terrace (12), and parking (13).

In this study, the minimum area requirement for on-site medical care is mapped in the thirteen studied areas of the three-building case studies. The area sufficiently supporting the on-site medical service is marked and compared. The obstacles, obstruction, and barriers within the space for the medical service are then identified and proposed and noted as references for improvement in future projects

Rescue equipment	Dimension (m^2)			
	1 Equipment sizing	2 Human scale	3 Activity	4 Area requirement
Long spinal board				
	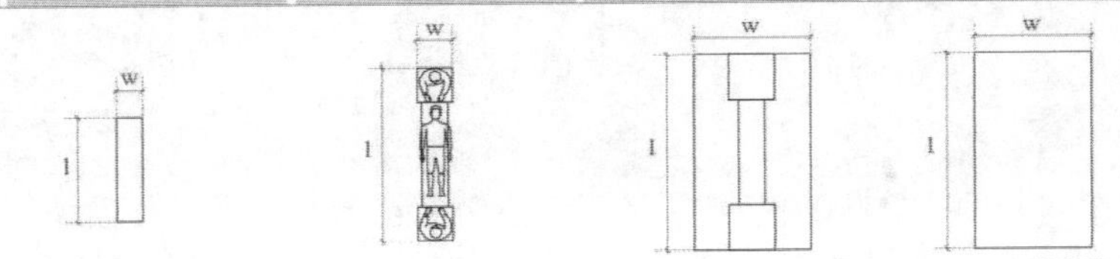			
area				7.062 m^2
	l w h	l w h	l w h	l w h
	1.845 0.450 0.050	3.045 0.600 1.710	3.445 2.050 1.710	3.445 2.050 1.710
Emergency bag and accessories				

area				0.467 m^2
	l w h	l w h	l w h	l w h
	0.356 0.229 0.229	0.600 0.829 1.710	0.600 0.829 1.710	0.600 0.829 1.710

*l = length, w = width, h = height

Figure 1. The size of equipment, dimension and area requirement for the medical care activities.

3. Results and Discussion

Figure 2 presents the mapped house's floor plan with the area requirement for on-site medical care. The rectangular shape with the red highlight represents insufficient space for on-site medical care, while those with the green highlight can support the medical service (figure 2(1)). Figure 2(2) shows the locations and types of barrier for the medical service in the studied areas. Overall, the area of the kitchen, bedroom, stair, and toilet is relatively small. Even the area is larger than that requirement, the perimeter of this area is inadequate for that operation. The walls are a significant barrier for taking care of such small rooms, whereas the obstruction for the on-site medical care in dining and living areas mostly area the arrangement of heavy and built-in furniture. In House A, it is found that the obstacles include 17 furnitures (F1, marked as the yellow circle) (mobile furniture such as chairs is not counted), 11 indoor walls (W1, marked as the red circle), and 14 outdoor walls (W2, marked as the green circle). For House B, the obstacles comprise 16 furniture (F1), 11 indoor walls (W1), and 17 outdoor walls (W2). House C has many obstacles but fewer than those found in House A and B, in which the obstacles include 15 furniture (F1), 9 indoor walls (W1), and 16 outdoor walls (W2).

Table 1 presents a score calculation of available areas that can corresponds to the on-site medical care weighted by thirteen studied areas. Overall, all studied houses have available space supporting on-site medical care for only 2 areas from a total of 13 areas, which accounts for 15 percent. One of those areas found in all three houses is car parking. The interior space is adequate for the medical care for House A and B, then it is the corridor and dining area, respectively. For House C, all available areas are outdoors, including car parking and outdoor terrace.

In comparison to adequate space for medical care and first aid in the three studied houses. It is found that House A and B is stronger in readiness for emergency situations than that of House C. It is because these two houses have medical care area in interior space, whereas both spaces of House C are located outdoor. Overall, the design of room layout and areas of the typical single houses in Thailand are not sufficiently supported the on-site emergency medical service. These barriers could delay access to medical treatment for older adult's injuries on site, which leads to serious health effects and reduces the quality of life.

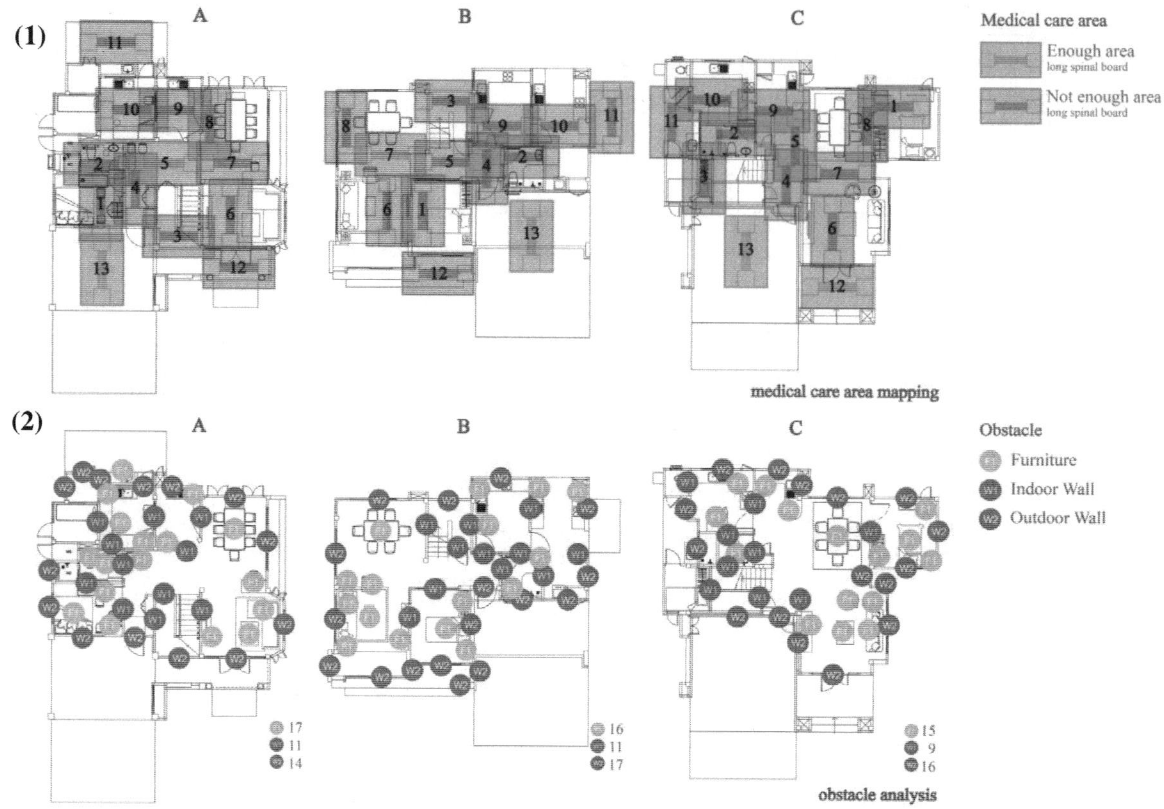

Figure 2. (1) Presents the mapping of medical care areas in different locations on the 1st floor of three houses for this case study. (2) Depicts the obstacles and barriers for medical care.

Table 1. Available the medical care area scoring

No.	**Area**	**Available area supporting the on-site medical care**								
		House A			**House B**			**House C**		
		$m^{2,a}$	$\%^b$	Score	$m^{2,a}$	$\%^b$	Score	$m^{2,a}$	$\%^b$	Score
Indoor										
1	Bedroom	2.71	38%	0	4.31	61%	0	4.38	62%	0
2	Toilet	2.21	31%	0	2.25	32%	0	2.32	33%	0
3	Stair	2.11	30%	0	2.07	29%	0	2.12	30%	0
4	Corridor 1	4.32	61%	0	2.54	36%	0	4.59	65%	0
5	Corridor 2	7.06	100%	1	5.50	78%	0	7.05	99.8%	0
6	Living area	6.11	87%	0	5.85	83%	0	5.85	83%	0
7	Dining area 1	6.30	89%	0	7.06	100%	1	6.43	91%	0
8	Dining area 2	3.79	54%	0	4.65	66%	0	3.99	56%	0
9	Pantry	3.49	49%	0	3.83	54%	0	4.68	66%	0
10	Kitchen	3.55	50%	0	3.80	54%	0	4.53	64%	0
Outdoor										
11	Washing area	6.89	98%	0	3.83	54%	0	3.77	53%	0
12	Terrace	4.75	67%	0	3.41	48%	0	7.06	100%	1
13	Car parking	7.06	100%	1	7.06	100%	1	7.06	100%	1
Total indoor pts.				1 (8%)			1 (8%)			0 (0%)
Total outdoor pts.				1 (8%)			1 (8%)			2 (15%)
Total pts.				2 (15%)			2 (15%)			2 (15%)

[a] is an available area (empty space).
[b] is a percentage of available area weight by the minimum area requirement for on-site medical care.

4. Conclusion

This study evaluates the design of room layout and areas of typical single houses in Thailand that can support Emergency Medical Services to aid the elderly having fall injuries in the house. The on-site medical service and evacuation devices are gathered by the interview of medical specialists and rescue teams. These collected data are used to calculate the area required for the medical treatment operated by three medical staff proceeding their tasks along the perimeter of the long spinal board. Such a minimum area requirement is then mapped and evaluated with the floor plan of three single houses, commonly found in neighborhood districts in Bangkok. It is found that the designed layout and room area of all studied houses insufficiently support the medical treatment on site, which has only two available areas. One of those is outdoor car parking. Considering safety for indoor living environment, House A and B provide more comfortable space (founded in the dining area and indoor corridor) for on-site medical treatment than that of House C. For future studies, the flexible design for interior walls and the arrangement of mobile and innovative furniture in the interior of residences providing the medical care in the houses should be further explored. Emergency Medical Services underlies several processes, in which one significant part is an evacuation of the patient into the ambulance. Instead of the medical treatment on site, future studies should examine whether the space allowing circulation and evacuation permits rapid movements in the urgent case of necessary transportation the seniors in their houses. Ultimately, these studies serve as crucial guidance in reducing the risk of fall injuries in the elderly in the houses.

5. References

[1] World Health Organization 2008 WHO Global Report on Falls Prevention in Older Age. Available from: http://www.who.int/ageing/publications/Falls_prevention7March.pdf

[2] Huang T 1998 Fall-Prevention in Taiwanese Elderly Adults (Order No. 9937045). *Available from ProQuest Dissertations & Theses Global* (304449818). Available from: https://search-proquest-com.ezproxy.tulibs.net/docview/304449818?accountid=42455

[3] Kotaki A 2013 The Impact of the Interior Environment on Resident Falls in Japanese Nursing Homes (Order No. 1524232). *Available from ProQuest Dissertations & Theses Global.* (1460178564). Available from: https://search-proquest-com.ezproxy.tulibs.net/docview/1460178564?accountid=42455

[4] Zhao D, Yang J, Wang Y, Yu C, Jiang Y, Wang S, and Fukuda T 2018 Multiple Welfare-Robots Architecture of Smart House for Security Assistance of Daily Activities. *2018 IEEE International Conference on Intelligence and Safety for Robotics (ISR)*, art. no. 8535715: **595-601**.

[5] Halvorsen H-P, Kaspersen K, Jonsaas A, and Mylvaganam S 2019 Fusion of Body Sensors' Data and Video Images in Assistive Technology. *2019 IEEE Sensors Applications Symposium (SAS),* art. no. 8706053: **1-6**.

[6] Faul M, Stevens JA, Sasser SM, Alee L, Deokar AJ, Kuhls DA, and Burke PA 2016 Older Adult Falls Seen by Emergency Medical Service Providers: A Prevention Opportunity. *American Journal of Preventive Medicine*. 50(6): **719-726**.

[7] Jareemit D, Inprom N, and Sukseeda J 2016 Uncertainty Distributions in Architectural Design Parameters for Destached Houses Located in Bangkok Neighborhoods. *Proceedings of SimBuild*. 6(1): **197-204**.

Acknowledgment

We are deeply grateful for special support from Thammasat University and NSTDA for Excellent Research Graduate Scholarship, Thammasat Design School for the International conference scholarship, Dr. Sattha Riyapan (Head of EMS Unit of Siriraj Hospital), and Mr. Phubet Ploysawat (Head of EMS Unit of Chulabhorn Hospital) for providing EMS information and recommendations.

IOP Conf. Series: Materials Science and Engineering **910** (2020) 012018 doi:10.1088/1757-899X/910/1/012018

Factor Analysis of Worker Engagement in Construction Projects in Thailand

Grit Ngowtanasuwan

Construction Management Program, Faculty of Architecture, Urban Design and Creative Arts, Mahasarakham University 44150 Thailand, grit_n@hotmail.com (+66) 818712414

grit_n@hotmail.com

Abstract. Factor analysis of worker engagement in Thai construction projects was undertaken. Exploratory factor analysis (EFA) was conducted on a sample group of 210 construction project workers in the study area of the northeastern central region of Thailand. Results indicated six factors of worker engagement as (1) relationship and dedication, (2) workplace, (3) part of the company, (4) expertise, (5) pride in work, and (6) relationship with supervisors. Strategies for strengthening worker engagement in construction projects are presented in the conclusions.

1. Introduction

In Thailand, rapid population increase along with industrial and social growth have resulted in the construction industry becoming the beating heart of economic development [1]. Demand for construction labor has increased with the establishment of the ASEAN economic community, allowing provinces in the region to experience rapid economic growth with high investment in infrastructure and building construction. Provinces in the northeastern central region of Thailand as Roi Et, Khon Kaen, Maha Sarakham and Kalasin are continuously expanding in population, economy and the society. The construction business is a service industry where contractors emphasize the quality of their work to satisfy project managers and compete for engagement on future projects [2]. The construction industry in Thailand is highly competitive in terms of cost, quality of work and duration of construction. To be successful, a construction project requires the services of skilled labor to positively engage with the organization or company and generate high- quality work to complete the project on time. A contractor is responsible for providing all the materials, labor, equipment, and services necessary for the construction of a project [3]. To sustain business in the industry, contractors need to know and understand the factors relating to maximizing worker engagement and the optimal strategies to be followed. Hence, this study analyzed the factors and measurement items related to worker engagement in the Thai construction industry. Results will allow contractors to retain and engage their workers within the industry. The scope of this study is based on an opinion survey in terms of engagement of construction project workers in the study area of the northeastern central region of Thailand consisting of Roi Et, Khon Kaen, Maha Sarakham and Kalasin Provinces. Strategies for the development and support of worker engagement in construction projects in Thailand are presented.

2. Research background

The Cambridge Dictionary [4] defines a contractor as a person or company that arranges to supply materials or workers for building construction and performs work on a contractual basis. A contractor is responsible for providing all materials, labor, equipment, and services necessary for the construction of a project. A general contractor often hires specialized subcontractors to perform all or parts of the construction work [3]. When using subcontractors, the general contractor is responsible for the quality of all work performed by any and all of the hired workers [3]. Employee engagement had been defined by many reputable research institutions such as Gallup Strategic Consulting [5], Mercer Consulting Group [6], Hay Group [7] and Aon Hewitt [8]. To summarize, employee engagement involves a positive attitude with confidence and faith in the organization through accumulation and continuous development. Employees must feel that they are a part of the organization and want the company mission to succeed. Employee engagement is an expression of behaviors (3S: Say-Stay-Strive) [8]. (1) Employees speak (Say) positive things about the organization, (2) from their existence (Stay) employees desire to continue to be members of the organization, and (3) employees use their best efforts (Strive) to help or support the success of the organization. Employee engagement is not a spontaneous asset but something that must be created through proactive activities. Employee engagement reduces staff turnover and this increases revenue and profit [9]. In the construction industry, employee engagement is termed as worker engagement and the employer is referred to as the contractor. Worker engagement can be measured by many factors including pride in work, part of the organization, desire to see the organization succeed, relations with and acceptance by colleagues and supervisors, and nature of the workplace [5], [6], [7], [8], [9].

3. Research methods

3.1. Identification of the factors

The Delphi technique is a widely used and accepted method for achieving convergence of opinion concerning real-world knowledge solicited from experts within certain topic areas [10]. This technique is designed as a group communication process of experts that aims at conducting detailed examinations and discussions of a specific issue. The process can be continuously iterated until a consensus is determined to have been achieved. This technique was used to confirm the factors and the conceptual research model. The Delphi technique began by identifying the factors, related theories and literature available as textbooks and research articles. All the factors were summarised, listed and then confirmed by the experts using the Delphi technique. A group of five experts in related fields was chosen based on their experience and included two construction project managers with more than 20 years of experience in the Thai construction industry, two experts from construction consultancy companies, and one expert from an academic institution. The listed factors were confirmed by a consensus of opinions of the five experts both directly and individually starting from expert 1, 2 ..., 5 and returning to expert 1 circularly. All experts carefully and individually considered the listed factors without meeting each other for at least two times per expert until the consensus was saturated. Once the listed factors were confirmed and categorized by the experts they were constructed and made ready for analysis. Initial factors and items are listed in Table 1.

3.2. Questionnaire design

Data collection was carried out using a questionnaire survey to determine the principal causal factors leading to Thai construction workers. The questionnaire was subdivided into two factor categories with the items selected as listed in Table 1. The initial section covered demographic data including gender, age, marital status, work experience, education, daily income, project location, project value, and project duration (9 items). The second section covered organizational engagement (12 items), while the third section covered personal engagement (10 items). Demographic data were presented by frequency and percentage, while the other sections asked respondents to answer questions using a 5-point Likert scale from ‘strongly disagree’ to ‘strongly agree’.

Table 1. Initial factors and items

Factor	Item
Organizational engagement	WE1: Pride in the profession
	WE2: Being accepted by colleagues and helping each other well
	WE3: Being entrusted to take responsibility for duties that match knowledge and expertise
	WE4: Freedom to work and be able to offer opinions about work responsibility
	WE5: Able to work with supervisors and understand each other well
	WE6: Supervisors always advise and give you support
	WE7: Able to work well with colleagues
	WE8: You and your colleagues have good relations with each other, there is regular cooperation and assistance
	WE9: The company/contractor has work for you to do consistently and continuously
	WE10: Having the opportunity to be promoted
	WE11: The workplace is clean and tidy
	WE12: Modern work tools
Personal engagement	WE13: Ready to dedicate and sacrifice to the company/contractor
	WE14: Ready to cooperate with the company/contractor without any doubts
	WE15: Proud to be part of this company/contractor
	WE16: The problem of the company/contractor is like your problem
	WE17: This company/contractor is like your home
	WE18: Higher wages cannot make you decide to change jobs
	WE19: It is wrong to decide to leave this company/contractor while the company/contractor is experiencing problems
	WE20: You and your colleagues have a good relationship with each other, have regular cooperation and help each other
	WE21: Proud to tell others that you work here
	WE22: When you work, you try your best to do more than you can

3.3. Validity and reliability tests

Validity and reliability tests were employed to ensure that the questionnaire items were suitable for their purpose. Validity testing involved interviews with five experts to discuss the factors identified in the previous section. The expert reviewers confirmed that the items selected were appropriate for the measurement of the study, while also making further suggestions for items that might improve the validity in this particular context. Validity testing is an important step to ensure that the items are valid while avoiding confusion and ambiguity. Cronbach's alpha was used to analyze the reliability of the items and assess the responses in a pilot study involving 30 participants who were Thai construction workers and matched the study population. Twenty-two questionnaire items, measured by a Likert scale (12+10 items of organizational and personal engagement categories, respectively) were analyzed using statistical software. Overall, Cronbach's alpha coefficient for the 22 items was 0.731, and it was observed that the result would be 0.736 without the presence of item WE9. Accordingly, WE9 was removed from the study. For the organizational and personal engagement categories, the Cronbach's alpha coefficients were 0.722 and 0.727, respectively. Since none of the coefficients fell below 0.7, the questionnaire was deemed as reliable [11].

3.4. Data Collection

After completing the questionnaire design, the next stage involved selection of the sample group of Thai construction workers using a convenience non-probability sampling approach. The respondents were predominantly based in the study area of the northeastern central region of Thailand consisting of Roi Et, Khon Kaen, Maha Sarakham and Kalasin Provinces. The survey was carried out over a three-month period. Interviews were conducted face-to-face to ensure that all the respondents fully understood the

IOP Conf. Series: Materials Science and Engineering **910** (2020) 012018 doi:10.1088/1757-899X/910/1/012018

purposes of the research. A total of 300 questionnaires were distributed, and 210 were accepted as valid and processed.

Table 2. Descriptive results

Description	Frequency	Percentage
Gender		
- Male	109	51.9
- Female	101	48.1
Age (years)		
- Less than 30	23	11.0
- 40 - 31	58	27.6
- 50 - 41	74	35.2
- More than 51	55	26.2
Marital status		
- Single	73	34.8
- Married	114	54.3
- Widow/divorce	23	11.0
Work experience (years)		
- Less than 1	75	7.35
- 1 - 5	78	1.37
- 6 - 10	30	3.14
- 11 - 15	11	2.5
- 16 - 20	10	8.4
- More than 21	6	9.2
Education		
- Primary school	129	61.4
- Secondary school	73	34.8
- High school	2	1.0
- Vocational certificate	6	2.9
Daily income (Baht)		
- Less than 300	41	19.5
- 300 - 350	132	62.9
- 351 - 400	27	12.9
- 401 - 500	10	4.8
- More than 500	0	0.0
Project location		
- Roi Et	50	23.8
- Khon Kaen	50	23.8
- Maha Sarakham	60	28.6
- Kalasin	50	23.8
Project value (Baht)		
- Less than 10 M	60	28.6
- 10 - 100 M	30	14.3
- More than 100 M	120	57.1
Project Duration (years)		
- Less than 1	91	43.3
- 1 - 2	119	56.7
- More than 2	0	0.0

4. Results

4.1. Descriptive results

The 210 questionnaires completed by construction project workers in the study area were analyzed. Descriptive results are shown in Table 2

4.2. Exploratory factor analysis
Exploratory factor analysis (EFA) is a statistical method for uncovering the underlying structure of a relatively large set of variables [12]. EFA is used to reduce the number of variables to a smaller set of underlying summary variables called factors. In this research, EFA was implemented to determine the underlying structure of factors related to the worker engagement and implemented with varimax rotation using computer statistical software. The output showed the Kaiser-Meyer-Olkin (KMO) measure of sampling adequacy at .647 (KMO>0.6) [13]. Bartlett's test of sphericity recorded a significant value of .000 (less than .05) with an approximate chi-square of 760.898, df = 171. Factor loading values less than 0.4 were eliminated. The extraction of EFA was based on initial eigenvalues greater than 1 [14] (Table 3), and the output analysis indicated six components as the number of factors in this EFA. The percentage of explained variance was 67.06%.

Table 3. Factor loading and Cronbach's alpha

Item	Factor						Cronbach's alpha	
	1	2	3	4	5	6		
WE20	789.0	-100.0	139.0	047.0	122.0	-120.0	0.720	0.722
WE22	721.0	031.0	018.0	123.0	-110.0	177.0		
WE19	665.0	361.0	109.0	-118.0	-028.0	035.0		
WE15	592.0	048.0	100.0	-025.0	160.0	323.0		
WE13	528.0	287.0	-174.0	122.0	114.0	-229.0		
WE11	091.0	792.0	074.0	132.0	099.0	003.0	0.706	
WE12	089.0	690.0	104.0	080.0	010.0	237.0		
WE16	-072.0	101.0	753.0	041.0	042.0	107.0	0.764	
WE18	029.0	276.0	726.0	-057.0	086.0	-334.0		
WE21	371.0	-229.0	577.0	097.0	-090.0	187.0		
WE14	342.0	-003.0	415.0	-016.0	077.0	069.0		
WE3	041.0	042.0	-006.0	788.0	-016.0	-159.0	0.744	
WE10	063.0	422.0	028.0	628.0	-150.0	057.0		
WE4	016.0	-023.0	119.0	597.0	341.0	361.0		
WE1	118.0	061.0	072.0	-122.0	738.0	036.0	0.715	
WE7	-010.0	-059.0	064.0	052.0	735.0	055.0		
WE2	066.0	304.0	-145.0	368.0	513.0	-076.0		
WE5	097.0	069.0	110.0	-161.0	194.0	699.0	0.726	
WE6	060.0	283.0	-107.0	222.0	-208.0	538.0		

The output from EFA indicated 19 remaining items after WE8 and WE17 were removed from the analysis. The items were categorized into six groups as factor 1 (WE20, WE22, WE19, WE15 and WE13), factor 2 (WE11 and WE12), factor 3 (WE16, WE18, WE21 and WE14), factor 4 (WE3, WE10 and WE4), factor 5 (WE1, WE7 and WE2), and factor 6 (WE5 and WE6). For these factors, the average values of Cronbach's alpha were 0.720, 0.706, 0.764, 0.744, 0.715 and 0.726, respectively. Nunally [11] considered that a Cronbach's alpha value greater than 0.7 could be deemed acceptable. Therefore, these factors were acceptable with overall Cronbach's alpha value for the six factors at 0.722.

5. Conclusions

Exploratory factor analysis was conducted in the context of engagement among Thai construction workers. A total of 210 sample questionnaires were completed by respondents who worked on construction projects in the study area of the northeastern central region of Thailand consisting of Roi Et, Khon Kaen, Maha Sarakham and Kalasin Provinces. Data were analyzed to confirm the relevant factors. Demographic profiles of the participants are shown in Table 2. EFA was used to identify and confirm the relevant factors that influenced the engagement of Thai construction workers. Six factors were identified by EFA as 5 items of factor 1, 2 items of factor 2, 4 items of factor 3, 3 items of factor 4, 3 items of factor 5 and 2 items of factor 6. These were used to make recommendations concerning worker engagement in construction projects in Thailand. The two highest (1st and 2nd items) from the factor loading values (shown in Table 3) were chosen to develop the guidelines. Each factor was named by implication of their measurement items. The six factors of worker engagement in construction projects in Thailand are presented in Table 4.

Table 4. Factors of worker engagement in construction projects in Thailand

Worker Engagement	Relationship and dedication	• You and your colleagues have good relations with each other, there is regular cooperation and assistance • Ready to dedicate and sacrifice to the company/contractor
	Workplace	• The workplace is clean and tidy • Modern work tools
	Part of the company	• The problem of the company/contractor is like your problem • Higher wages cannot make you decide to change jobs
	Expertise	• Being entrusted to take responsibility for duties that match knowledge and expertise • Having the opportunity to be promoted
	Pride in work	• Pride in the profession • Able to work well with colleagues
	Relationship with supervisors	• Able to work with supervisors and understand each other well • Supervisors always advise and give you support

The factors and items listed in Table 4 can be used to formulate strategies to promote engagement among Thai construction workers and create competitive advantage in the Thai construction industry.

6. Reference

[1] Krungsri 2020 [cited 2020 Mar 1] Available from https://www.krungsri.com/bank/getmedia/7fe1e975-13f4-4040-a276-0d40841e504b/IO_Construction_Contractor_190606_TH_EX.aspx

[2] Arslan G and Kivrak S 2008 Critical factor to company success in the construction industry, World Academy of Science, Engineering and Technology 45 (2008) pp 43-46

[3] Cambridge Dictionary 2020 Contractor cited 2020 Mar 1] Available from https://dictionary.cambridge.org/dictionary/english/contractor

[4] Wikipedia the free encyclopedia 2020 Contractor [cited 2020 Mar 1] Available from https://en.wikipedia.org/wiki/Contractor

[5] The Gallup Organization. 2008 Employee Engagement Gallup's Research-Based Approach. [cited 2020 Mar 1] Avaiable from http://gmj.gallup.com/content/126173/Bolster- Employees-Confidence.aspx

[6] Mercer Consulting Group 2007 Employee to drive global business success: insights from Mercer's What's Working research [cited 2020 Mar 1] Available from http://mmc.com/knowledgecenter/Engaging_Employee_To-Drive_Business.pdf

IOP Conf. Series: Materials Science and Engineering **910** (2020) 012018 doi:10.1088/1757-899X/910/1/012018

[7] Hay Group 2009 Engaging and Enabling Employee to improve Performance Outcomes [cited 2020 Mar 1] Avaiable from http://info-now.com/typo3conf/ext/p2wlib/pi1/press2web/html/userimg/EEA_EXPO_2010/Proceedings/SurveysandPolls-HayGroup-Agnew.pdf
[8] Aon Hewitt 2011 What is Employee Engagement [cited 2020 Mar 1] Available from https://ceplb03.hewitt.com/bestemployers/turkey/pages/emp_eng.htm
[9] Federman B 2009 Employee Engagement: A Roadmap for Creating Profits, Optimizing Performance, and Increasing Loyalty (San Francisco: Jossey-Bass A Wiley Imprint)
[10] Hsu C C and Sandford B A 2007 The Delphi Technique: Making Sense of Consensus Practical Assessment Research & Evaluation vol 12 (10)
[11] Nunnally J C 1978 Psychometric Theory 2nd edition (New York: McGraw)
[12] Norris M and Lecavalier L 2010 Evaluating the Use of Exploratory Factor Analysis in Developmental Disability Psychological Research Journal of Autism and Developmental Disorders Vol 40 (1) pp 8-20
[13] Tabachnik B G and Fidel L S 2001 Using multivariate statistics 4th edition (Massachusetts: Allyn & Bacon Needharm Hights)
[14] Kaiser K F 1958 The varimax criterion for analytic rotation in factor analysis Psychometrika vol 23 (3) pp 187-200

Acknowledgments
This research project was financially supported by Faculty of Architecture, Urban Design and Creative Arts, and Mahasarakham University Development Fund, Thailand.

IOP Conf. Series: Materials Science and Engineering **910** (2020) 012019 doi:10.1088/1757-899X/910/1/012019

Cost perspective for sustainability in Industrialised Building System (IBS) projects

M Z Mohammad[1], E F Ahmad[1] and N S Samsudin[2]

[1] Faculty of Civil Engineering, Universiti Teknologi MARA, Shah Alam 40450 Selangor, Malaysia.
[2] Faculty of Architecture, Planning and Surveying, Universiti Teknologi MARA Seri Iskandar Campus 32610 Perak, Malaysia.

mazlinazaira@uitm.edu.my

Abstract. Since 1998, the Construction Development Industry Berhad (CIDB) has been long promoting Industrialised Building System (IBS) in Malaysia. However, the use of IBS has not achieved an impressive number which only 10% and 50% have widely used in the past said CIDB in 2006. This research study focused on why IBS practice is still in an unconvincing market state of use and how sustainability in its cost perspective influences the use of IBS in each project. A questionnaire has been developed consists of four sections and has been distributed to 30 respondents in different IBS projects within the Selangor area in Malaysia. The data has been analysed using SPSS 23 software to determine the descriptive statistics and rank regarding the issues of implementing of IBS in the industry based on a cost perspective. The research study determined the level of understanding of IBS acceptance and awareness among the Malaysian construction industry. The findings encountered the main problems by constructions player while implementing IBS. In conclusion, strategies and recommendations are also suggested.

1. Introduction

Industrialised building system (IBS) is a term used by the local sector specifically Malaysia and the government to show ownership of adapting the industrialisation of building and prefabrication of components in building construction. IBS itself defined as one of the construction methods by means the components are assembled in controlled environment transported and structured into minimal site works [1,2]. Most of the manufacturing components are manufactured systematically using machine. While formworks and other forms of mechanical equipment are manufactured off-site and delivered to site for assembly and erection. Minimal additional site work [3]. IBS have situated in Malaysia more than 50 years and many construction industries embrace IBS as their method of construction, realizing there are more of productivity, reducing risks related occupational safety and health, enhance unskilled and skilled worker's dependency on manual foreign labour. Other, IBS offer low wastage, lesser site materials and a neat and cleaner environment [1]. On top of that, conventional construction has been criticising for not fulfil the requirements as it includes a vast number of workers, human health and safety. Sustainability in construction is generally responsible for an efficient resource for building a healthy environment [4]. Now, Present environmental and social-economic problems revolve around sustainability. According to [5], IBS view as the most advantageous solutions to reduce waste in construction. On top of that, IBS is viewed as the best alternatives way to maintaining the sustainability in construction. It can supervise the human resources and cost, shorten the length of construction time and improve the quality of buildings. As compared to other first world countries such as Japan, Germany and United Kingdom has shown a great potential to progress for IBS to progress [6]. These are among the successful project IBS related projects Sekisui Home (Japan), Living Solution (United Kingdom), Open House (Sweden) and Wenswonen (Netherland). According to [7] to the unready state of a decision on choosing IBS over the conventional methods could be led to many other problems and the potential of the IBS unable to be optimised. Problems such as changing orders and delays in production, over budget encountered due to defective decisions in implementing IBS.

2. Literature review

2.1. IBS

According to [6], IBS technology is not new technology adopted in Malaysia since the Ministry of Housing and Local Government visited several European countries and assessed their building system. After that, the second IBS project was constructed at Jalan Rifle Range Penang in 1965, which consist of six block of 17-storey flats and three blocks 18-storey flats [8]. CIDB has been operatively promoting the use of IBS in every construction [9]. IBS is not a new technology. IBS can be defined in all of which every structure of the building such as a wall, slab, column, staircase is mass-produced in a factory or at a site factory undergo tight quality control and minimize the in-situ casting. There are several classifications of IBS modular structure. According to [3] IBS is divided into six components based on the structural aspects for example panel and box system, precast concrete framing, steel formwork systems, steel framing systems, framing timber systems, blockwork systems and innovative product system. Normally, it involves mass production of amount, ensures cost, and time reduction other than raising the quality of the work. The precast concrete components are widely used locally and abroad, and the components products are well manufactured and cured under some quality control from the plant. In addition, at some point the whereby the formwork works as a structure itself after concreting. Construction Sustainability means, alternately maintaining the environment while controlling human resources, cost, construction period and increase the quality of the building [7,10].

2.2. Advantages and disadvantages of IBS

IBS brings many benefits to the construction industry. According to [11] conventional on-site construction has long been criticized as it has long construction time, low productivity, high wastage, and low safety records. They are ways to reduce this one of them is implementing the IBS into every project related and start adopting IBS as a construction role in the industry. Despite there are many advantages of IBS implementation, there are also the disadvantages of IBS. [12] said the IBS manufacturing process needs early attention and a good plan from among all of the construction players. Such support is given by the government but still, usage of the IBS is currently much lower. IBS players do not find the right way or the right chemistry finding solution to challenges that indicates to supply chain of IBS [13]. Several Issues relates to IBS is most of the available workers are still lack of required skill and knowledge. Several workers available are hard to train and to attract labors [10]. On top of that, deficiency in skills and knowledge gave a visible impact to improper assembly of the components [8]. Other, IBS needs a large start-up capital cost, IBS needs investment for setting up the plant, machinery supply, and engineering complexity of the new interface and expenditures of transportation [14]. In addition, [15] stated project characteristics and the economic market are the common factors of IBS construction cost influencer. [10], a lack of knowledge is one of the main issues and disadvantages for IBS. Slow in terms of making a decision whether to choose IBS over the conventional methods to occur and the cost of it eventually hike up [1]. Lack of knowledge on IBS contributes to poor structural analysis and the design for prefabricated components [16].

3. Research methods

In this research, the sustainability aspect of IBS in term of cost in Selangor be stress specifically. This research purposely to conducted to gain knowledge of IBS implementations by construction players.

Data collection in this research is divided into two categories which is primary and secondary data. Data collected by the researcher is considered as primary data while other researcher collects the secondary data. The secondary data was in the form of articles journals or books. Other, the primary data we to collect with the help of construction players ranging from contractor engineers, manufacturers, clients and such. The questionnaire acquires the respondents to fill on the knowledge on the IBS project that are related, based on their experience before and opinion of the construction method implementation for future reference.

3.1. Data sampling and questionnaire formation

This study focuses on IBS projects and relating to the IBS all over around Selangor. The location of the study is chosen because of the amount of construction player Selangor shows the highest number according to CIDB. The case study refers to the implementation of IBS around Selangor and how the implementation of it cut costs and reduce the time of the construction project. This research study utilizes techniques of quantitative by gathering and collecting all information and data to fulfil the intended objectives. The quantitative research is used to give information in a graphical way. The procedures to collect data throughout this method usually by using a set of questionnaires with standardized measuring instruments, rating scales, and observation schedules. Under this method, the researcher uses the questionnaires to get the variables required or known as primary information to be used to perform the analysis of data. The questionnaire is distributed to qualified respondents' expert to the IBS construction method, sustainability approach, cost and time management skill. A set of questionnaires was formed and subsequently distributed to the respondents intended to solicit reactions and their opinions on the cost sustainability in IBS. The author's research was conducting a "pilot survey" over the study area to ensure that questionnaires be distributed to achieve the objectives of the research that have been set. The questionnaire is divided into four main sections includes Section A to evaluates the demographic information. Likert scale method be used to perform the analysis for the following sections. Section B is to answer on the benefit of implementations of IBS. The respondents need to choose an answer from strongly agree to strongly disagree for each point related to the project. There are ten questions prepared in this section to evaluate the satisfaction feedback on services in Section C. Lastly is Section D questions related to the challenges occurred in managing an industrialized building construction project. In this research, a questionnaire was made online by using Google Form. In addition, as a precaution, the questionnaire is printed out as a form of hard copy and distributed physically to the respondent to increase a number of responds.

4. Results and discussion

Majority of the respondents are working with consultant and contractor companies, there are about 40% of the respondent comes from contractors. Thirty-three percent of the respondent is from consultants and only 27% comes from client or developer background. About 37% have worked in the field for about 5-10 years. Ten percent of the respondent 10 -15 years of experience and the rest at 7% have work for more than 15 years. About 73% of the respondent did have experience in the past with IBS project. While, about 20% of the respondent have not involve themselves in any experience before. Only 7% of the respondent is not sure about having relate to any IBS project before.

4.1. Mean and ranking for each section

According to **Table 1**, the analysis can be concluded on the benefits of implementing IBS in daily construction work. Data shown above depicts the value of standard deviation and mean for each question where most of the result is then analysed with the aid of SPSS 23 software. **Table 2** shown the result on ranking the issues related on cost in implementing of IBS in the construction industry. Justification on each result has been discussing in the following subsection.

4.2. Findings discussion

According to the result, the benefits of IBS, many of the respondents agreed on IBS use less material on site. [17] mentioned less material, less wastage, less volume, and speed up construction can be achieved. The result shows the highest mean on the question, this shows that respondents understand the most to this question as IBS usage may reduce the number of materials on site. Off-site production creates a more sustainable development for the construction industry to maintain the market value. A recommendation to aid future study should be made to increase understanding of IBS can be made. As mentioned in the Literature review, off-site production like IBS panel are manufactured off-site, this shows less material transported into the site and ease the site operation. Site operation is lesser, gain the second most response. This is proven by a previous researcher, [6] mentioned operation on site can be lessen as many of the IBS products are manufactured off-site. The statements are also supported by [6], construction on site can only be minimized and effectively handle by IBS implementation.

Table 1. Descriptive statistics and rank that aims cost perspective beneficial value of IBS

Description	N	Min	Max	Mean	Rank	Standard deviation
IBS uses less material as it is manufacture in factory	30	2	5	4.1668	1	.79145
IBS site operation is lesser	30	2	5	4.1333	2	.73030
IBS produced less wastage	30	3	5	4.1000	3	.60743
IBS keeps the site neat and tidy	30	3	5	4.1000	4	.60743
IBS promotes productivity(safe) and lessen the hazard	30	2	5	4.0667	5	.69149
IBS construction reduce number of workers on site	30	2	5	4.0333	6	.80872
IBS is a quick construction method	30	1	5	4.0000	7	.98261
IBS utilized minimal skill worker and machinery	30	1	5	3.6333	8	1.18855
Transportation cost and time is saved	30	1	5	3.6333	9	1.12903
IBS improve human daily skill and increase advancement capacity	30	1	5	3.6000	10	.96847

Table 2. Descriptive statistics and rank regarding cost perspective issues of implementation of IBS in the industry

Description	N	Min	Max	Mean	Rank	Standard deviation
Large scale productivity needs huge amount investment	30	3.00	5.00	4.3000	1	.59596
Start-up Capital cost is higher to work on	30	3.00	5.00	4.1667	2	.53067
Issues on training the labor	30	3.00	5.00	4.1667	3	.79148
Every parties sets different objectives	30	3.00	5.00	4.1000	4	.71197
Lacks of experience professional	30	1.00	5.00	3.9667	5	.80872
IBS is avoided due to lack of knowledge	30	2.00	5.00	3.7667	6	.81720
Unaware of the benefits of implementing IBS	30	1.00	5.00	3.7000	7	.98786
Lacks of guidance and different approach IBS are hard to be implemented	30	1.00	5.00	3.6333	8	.80872

As suggest by CIDB to increase the number of IBS implementation on site. IBS can be cost-efficient by only adapting the locking brick method. The respondents correspond the most to this as most of them are from the construction player themselves. However, shows the awareness of adapting IBS into the current construction method. To meet expectations project delivery without delay is achievable. [11] mentioned wastage produced between conventional and IBS method has been long criticized. The respondent is well agreed on IBS produce less waste and made the site cleaner. Waste on-site relates to the cost by transporting out the accumulating construction waste on-site. Waste such as resource and material waste on conventional methods is inevitable, but not on IBS adoption as supported by [18], IBS offers the most advantageous solution when it comes to waste. Less percentage of waste shows a cleaner and a tidy site. On top of that, recommendations on future studies can be made on IBS can improve human skills. Based on previous research, [5] suggest that IBS improves capacity and enhancement and quality of human life.

Referring to **Table 2** result, the results shows with a highest mean value of them to IBS needs a huge start-up capital. As mentioned by [19], Cost starting up for IBS project is undoubtedly high and become the most concerns for every construction-related party. Normally IBS involves with mass production, mass production means a bigger company to start-up with higher cost of money. IBS needs a bigger investment unlike small start-up company cannot contest and participate in any IBS related projects.

IOP Conf. Series: Materials Science and Engineering **910** (2020) 012019 doi:10.1088/1757-899X/910/1/012019

Although, the response from the questionnaire shows the higher mean value on the complication of training the labor. As many of the labor are well conversed in the conventional method of construction. [20] said challenges were encountered during the project phase whereas the labor was stuck with conventional minded method instead of the new IBS method. Management issues will then follow up as supported by [19], the manager has to ensure the workers well understand the method of the construction to accomplish the project and ensure its success. A small portion as only 30 respondents were actively participating in to answer the questionnaire Selangor. In addition, this shows a glimpse of the current construction player perception on the benefits and issues of IBS implementation, however, reflects the cost reliant on IBS implementation as many of the construction player's contradictory with each other.

5. Conclusion

This research has recognised the sustainability of IBS on its cost perspective. It can be concluded that the benefits of the IBS can be obtained with the study helps the construction industry to bloom in the future. However, not all parties could accept IBS implementation especially the company and individual with less experience in managing the IBS. The study has shown there was not a satisfied level of understanding for Malaysian industry members. On top of that, the awareness and acceptance of IBS should be strengthened and enforce in order for the IBS to use wide in Malaysia. The study has shown there was not a satisfied level of understanding for Malaysian industry stakeholders.

6. References

[1] Hamid Z A and Kamar K A M 2012 *J. Cons Innovation: Info, Process, Manag,* **12** 10.

[2] CIDB 2007 *Construction industry master plan CIMP 2006-2010 Construction Industry Development Board (CIDB)* (Malaysia, Kuala Lumpur)

[3] CIDB 2010 *Roadmap for Industrialised Building System (IBS) in Malaysia 2011-2015* (Malaysia)

[4] Kibert C J 2007 *J. Build Research & Info* **35** 601

[5] Baldwin A C S, Poon L Y, Shen S, Austin and Wong I 2009 *J. Renew. Energ.* **34** 2073

[6] Thanoon P, Lee W, Kadir, Mohd R A, Jaafar, Mohd S and Sapuan M 2003 The essential characteristics of industrialised building system. *Paper presented at the International Conference on Industrialised Building System, Kuala Lumpur, Malaysia*

[7] Luo Y, Riley R, Horman M J and Kremer G O 2008 Decision Support Methodology for Prefabrication Decisions on Green Building Projects. *Symposium on Sustainability and Value Through Construction Procurement, 29 November – 2 December 2006 University of Salford.*

[8] Kamar K A M, Hamid Z A, Ghani M K and Rahim A H 2007 Industrilized Building Systems: Current Shortcomings and the Vital Role of R&D 2nd (Quarter Master Building Publication)

[9] Mohamad M I, Zawawi M and Nekooie M A 2009 *J. of Civ. Eng.* **21** 234.

[10] Blismas N and Wakefield R 2008 Offsite manufacture in Australia - barriers and opportunities. *Paper presented at the Third International Conference of the Cooperative Research Centre (CRC) for Construction Innovation.*

[11] Eastman C M 2008 *J. of Cons. Eng. and Manag* **134** 526

[12] Pan, Wei, Gibb, A G F and Dainty A R J 2008 *J. Build. Res. & Info.* **36** 67

[13] Mohammad M F, Saifuza A, Shukor A, Mahbub R and Halil F M 2014 *Procedia - Social and Behavioral Sciences*, **153** 54

[14] Onyeizu E N and Abu Bakar A H 2011 *J. of Acade. Res.* **3** 175.

[15] Elhag T M S, Boussabaine A H and Ballal T M A 2005 *Int. J. of Proj. Manage.* **23** 545.

[16] Rahman N A A, Ahmad S and Zainordin Z 2013 *Procedia Social and Behavioral Science* **85** 69.

[17] Kamar, Kamarul A M, Hamid Z A, Mohd Khairolden, Zin M Z M, Rahim A H and Karim A Z A 2010 The critical success factors for the implementation of industrialised building in Malaysia. *Paper presented at the 3rd IBS Roundable Workshop (IRW03)-CIDB/CREAM IBS Survey.*

[18] Villa H, Resort B, Jabar I, Ismail F and Mustafa A A 2013 *Procedia - Social and Behavioral Sciences*, **101** 89

[19] Memon A H, Abdul Rahman I, Abdullah M R and Abdul Azis A A 2010 *Int. J. of Sustain. Cons. Eng. & Tech* **1** 54

IOP Conf. Series: Materials Science and Engineering **910** (2020) 012020 doi:10.1088/1757-899X/910/1/012020

Recovery of public spaces for the conservation of green areas in Tablada Lurin

Luis. Ramos[1], Doris Esenarro[2], Ciro Rodriguez[3] and Jessica Lagos[4]

[1]Federico Villarreal University Lima, Lima, Peru
[2,3]Specialized Institute for Ecosystems and Natural Resources Research (INERN)
[4]Cesar Vallejo University / Lima, Lima, Peru

berto_ramos242@yahoo.es

Abstract. This research allows us to show common problems such as the green surface deficit, citizen insecurity, and lack of essential services that reflect social exclusion risks. To the above added the absence of infrastructure of the services provided by these systems such as recreation and recreation. Together, they are essential factors for the valuation of land use. Therefore, it is proposed to develop an analysis of recreational green areas appropriate to the needs of the city. With it, you will facilitate the realization of inventories that estimate social indicators, economic, and environmental indicators. The importance of planning recreational parks is to involve citizen participation as a necessary strategy to solve the problematic situations that arise; inclusive processes must be ensured, especially to meet the demands of spreading vulnerable social groups. It concludes the need to address green planning with a focus on economic, social, and ecological studies that encourage the involvement of society in this task.

1. Introduction

One of the fundamental considerations of this report is argued in the use and control of urban space in terms of green areas and recreation within the city, to give added value to deteriorated or depressed spaces for the inaction of municipal authorities and / or the laziness of the villages [1,2], If the city is created and molded by people, this is due to the activities that are generated in it and should be satisfied; Thus, in the green and recreational areas there is the physical imprint of the relations established between the habitants of a society, and it will be essential to identify and know the divided city and the social ties disrupted by the inequality [3], If we can facilitate the democratic and equitable control of the actors themselves over their public space, we can assist their intervention in the processes put in the value of the recreational green areas.

Is to contribute to guarantee peace, tranquillity and reduce crime and delinquency at the local level, through an adequate treatment of recreational green spaces to deliver to the enjoyment of the community [4], and therefore ward off children and youth from the ravages of not positively managing their free time and leisure, away from the vices that disrupt morality and good customs, which is ultimately part of the objectives of citizen security.

This is part of the democratization of public spaces and their accessibility without restrictions to the community.

2. Method

2.1 Type of investigation

A description will be made of the multidimensional factors and the different manifestations of phenomena associated with the use of recreational public spaces of Tablada de Lurin district, as to show figure 1, measuring their characteristics, observing the configuration and processes that make up these phenomena, which will be subjected to a physical analysis space. From this information, a diagnosis of the current situation will be prepared. That is, the facts are observed as presented in their real context and in a certain time or not and then analyzed. Therefore, in this design, a specific situation is not constructed, but the ones that exist are observed.

Likewise, bibliographic information on historical events will be collected, to give it a cultural identity and revalue land use [4].

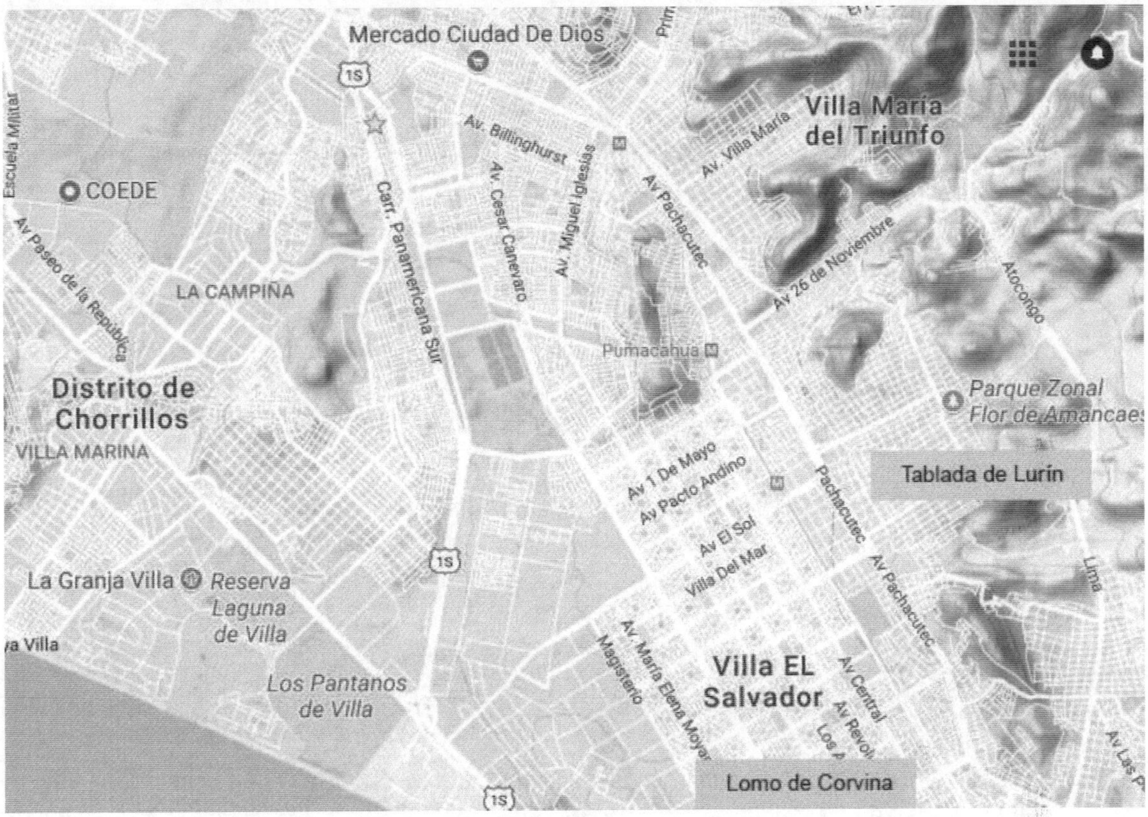

Figure 1. Plan of the fifth administrative area of Villa Maria del Triunfo district, Tablada de Lurin. Source: Megasoft

2.2 Procedures

These surveys were conducted in 04 Sectors of the Tablada de Lurin district, on different days and times where there was a higher concentration of public.

An estimate of costs per square meter of land in the 05 Sectors for the years 1995 and 2019. A survey conducted by information obtained from properties for sale.

3. Results

To Continuation, it shows the results obtained in the field survey on the land costs corresponding to the Plus dimension. According to Table 1, it is clearly established that the cost of land has increased to an average of 1384%.

Table 1. Increase in land cost according to sectors 1995 – 2019

	Cost/m2 (s/.)			
Sector	Year 1995	Year 2019	% Increase	% increase
1	67.80	866.67	798.87	1178%
2	40.68	780.00	739.32	1817%
3	36.16	650.00	613.84	1698%

4	12.05	162.50	150.45	1248%
5	9.04	97.50	88.46	979%

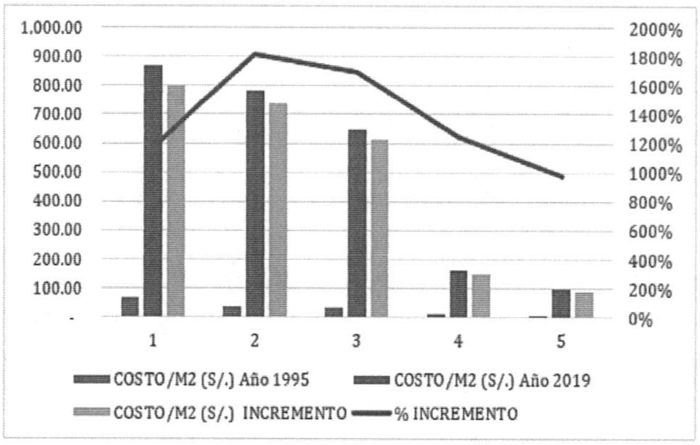

Figure 2. Increase of the land cost per m2(s /.), according to sectors.

According to Figure 2, in the curve of increase in the cost of land, it is observed in Sector 1 has the highest percentage of growth in these last 24 years; while Sectors 4 and 5 reflect a smaller, but significant increase with the investment made at the time of acquisition.

Table 2. Types of problems presented by the green recreational area.

	Frequency	Percentag e	Valid percentage	Accumulated percentage
Has no problems	9	7,5	7,5	7,5
Live people there	9	7,5	7,5	15,0
Is insecure	85	70,8	70,8	85,8
It is closed, there is no access	17	14,2	14,2	100,0
Total	120	100	100	

According to Table 2, there is a marked tendency in the area of Tablada de Lurin to insecurity with 70.8%, representing 85 respondents, while 14.2% refers to the inaccessible area. 7.5% of respondents report that their recreational green space does not present social problems. This is because the city is fenced and has access to certain times of the day.

On the other hand, 7.5% of respondents refer to the fact that the spaces are taken by antisocial persons (mainly drug addiction). The 14.17% of surveyed are not satisfied because the green recreational area is closed, and access is limited.

Table 3. Area reserved for privatized parks or recreation.

	Frequency	Percentage	% Valid	% Accumulated
Only park	86	71,7	71,7	71,7
Recreation only	17	14,2	14,2	85,8
Park and recreation	9	7,5	7,5	93,3
There are no reserved areas	8	6,7	6,7	100,0
Total	120	100	100	

According to Table 3, it is observed that the privatized public areas destined for parks are the most representative of the survey of 120 people.

Considering the areas destined for parks and recreation, 93.33% of the respondents are aware of some cases of loss of said areas for purposes beyond their purpose related to the urban plan. Land invasions for housing purposes [5].

Table 4 shows that public recreational areas have a weak general lighting system, according to a survey of 120 people.

Table 4. State of the public lighting service in green and / 21 or recreational areas.

	Frequency	Percentage	Valid percentage	Accumulated percentage
Bad	52	43,3	43,3	43,3
Regular	33	27,5	27,5	70,8
Good	34	28,3	28,3	99,2
Excellent	1	,8	,8	100,0
Total	120	100	100	

According to Table 4, it is observed that 43.3% of the respondents consider the public lighting service of the spaces understudy to be wrong, while 27.5% of them think it regular. 28.33% believe the lighting of park and recreation areas to be good, and 0.83% consider it acceptable. 29.16% of respondents report that the lighting quality is in the range of good to regular, while 70.84% rate it as consistent to bad. The green and recreational areas, with poor lighting, are located in the sectors away from the consolidated urban center, while those classified as useful to efficiently correspond to consolidated areas such as the Plaza de la Tablada de Lurin and the ecological park Flor de Amancaes [6].

Question 4: Do you have game equipment?

According to table 5, it is observed that the predominant equipment is excellent with more than five games and gym equipment for children and adults, represented by 46 respondents from a sample of 120 people.

Table 5. Game equipment.

	Frequency	Percentage	Valid percentage	Accumulated percentage
Bad (No games)	31	25,8	25,8	25,8
Regular (1-2 games)	4	3,3	3,3	29,2
Good (3-5 games)	39	32,5	32,5	61,7
Excellent (More than 5 games)	46	38,3	38,3	∞
Total	120	100	100	

Table 5 show 25.8% of the population considers that there is a lack of games in their sector, being the lousy qualifier, while 3.3% consider it regular where there are between 1 to 2 stationary games. 32.50% of the population think the state of conservation of sports slabs to be good, while 38.33% consider them excellent.

4. Discussion of results

The shortage of land to live, due to urban linear expansion, has increased the cost of the square meter of land throughout the metropolis, this is no stranger to the study area, encouraging invasions generally overlapping over green spaces. On the other hand, social insecurity (robberies, gangs, etc.), diminish the value of the land. Therefore, the tendency for family groups to improve their purchasing power, resolve to relocate to other residential areas of a better quality of services, there is an offer of transfers and/or sales of land and real estate. Converting green areas with a recreation component encourages neighbors to take care of the space and not allow invasion.

The inability on the part of the municipal administration to be able to offer essential services towards the public recreational areas has allowed areas to be taken by groups of people who live outside the law-making them insecure. On the other hand, the tendency to become wasteland conditions the possibility of being invaded by people who traffic with the land. The population claims to be taken care of by the environmental aspects of these areas [7], furthermore, the construction and maintenance of the infrastructure of green and recreational areas are part of the service that the municipality should be granted to the population since it helps the healthy recreation of the mind, body, and spirit of those who practice it. Away from the tremendous social problems identified and contribute to reducing crime rates and therefore protect children and youth, involving them in activities of enjoyment and coexistence.

5. Conclusions

The provision of services and infrastructure to these recreational public spaces allows revaluing the surrounding private lands, positively affecting what is the duty of the municipality to organize and train the neighbors about the management of their maintenance. Moreover, the evaluation of the services that the inhabitants demand of the green areas, evidences as an intrinsic necessity the recreation "the citizen security," whose causal relation: State of the recreational green regions and the insecurity, allows to argue that to foment the Control Natural of these public spaces is an instrument to prevent crimes and, therefore, mitigate vulnerability. Furthermore, to characterize a public space is to identify it with your neighborhood, which must use the essential services they demand, to impose its presence as a community. This inclusive social approach is incorporated through assessments that define the services that residents require from green areas as recreational and security needs [8].

6. References

[1] Abdulrahman Sa'adu Danjaji & Mariani Ariffin (2017). “Green infrastructure policy for sustainable urban development.” International article. *International Journal of Environment and Sustainable Development*, 2017 Vol.16 No.2, pp.112 – 127

[2] Maestre-Andres, S, Calvet-Mir, L, van den Bergh, JC (2016) Sociocultural valuation of ecosystem services to improve protected area management: a multi-method approach applied to Catalonia, Spain. *Regional Environmental Change* 16: 717–731

[3] Aguilar J. R (2018). A review of acoustic design criteria for school infrastructure in Chile. Journal Construction Engineering VOL.34 no.2. Santiago, CHILE. Available: http://dx.doi.org/10.4067/S0718-50732019000200115.

[4] Doris Esenarro, Ivana Escate, Leslie Anco, Carla Tassara, and Ciro Rodriguez (2020). Proposal for an Ecological Research Center for the Recovery and Revaluation of Biodiversity in the Town of Quichas-Lima, Peru. ¨ International Journal of Environmental Science and Development, ISSN 2010-0264, DOI: 10.18178, Vol. 11, No. 4. 212-216

[5] Jesus Gomez-Velazquez, Rosa Maria Velazquez-Sanchez, Abel Antonio Morales Santiago, and Armando Perez Delgadillo. Community Sustainability and Tourism in the Pooled Towns Route, Oaxaca, Mexico. *International Journal of Environmental Science and Development*, 2016. vol. 7, no. 2, pp. 156-159. Mexico.

[6] Anderson, CB, Pizarro, JC, Estevez, R, Sapoznikow, A, Pauchard, A, Barbosa, O, Moreira-Munoz, A, Valenzuela, AEJ (2015a) ¿ Estamos avanzando hacia una socio-ecologia? Reflexiones sobre la integración de las dimensiones ‘humanas’ en la ecología en el sur de América. Ecologia Austral 25: 263–272

[7] Ciro Rodriguez, Doris Esenarro, Pilar Ccorimanya, Fabiana Flores, Carmen Aylas & Jessica Lagos (2020). Proposal for a Sustainable Infrastructure Design (Ecolodge) in Quichas Town, Perú, ¨Test Engineering & Management¨ ISSN: 0193-4120 Page No. 9250 - 9256.

[8] Pascual, U, Balvanera, P, Diaz, S, Pataki, G, Roth, E, Stenseke, M, Watson, RT, Dessane, EB, Islar, M, Kelemen, E, Maris, V (2017) Valuing nature’s contributions to people: the IPBES approach. Current Opinion in Environmental Sustainability 26: 7–16

IOP Conf. Series: Materials Science and Engineering **910** (2020) 012021 doi:10.1088/1757-899X/910/1/012021

Theme park of renewable energies for mitigation of CO_2 in the urban area of the district of Chorrillos, Peru

Ciro Rodriguez[1], Doris Esenarro[2], Celso Alburquerque [3], Michelle Vega [4], and Valeri Ramirez[5]

[1,2]Pecialized Institute for Ecosystems and Natural Resources Research (INERN)
[3,4,5]Federico Villarreal University Lima, Lima, Peru

crodriguezro@unmsm.edu.pe

Abstract. The "Pantanos de Villa" is a natural habitat that conserves the high biodiversity of species and acts as the lung of the city, which is considered a wetland of international importance. This is being threatened by an inadequate infrastructure existing in the Lucchetti factory that attempts against the biodiversity of wetlands, for which the following project is proposed whose main objective is the mitigation of CO2 caused by polluting gases. Along with this, the integration of the neighbors with the swamps will be encouraged, through the design of an ecological and bio-sustainable park destined for recreational practices and sports, being a pleasant meeting point for families and tourists. With the theme park of clean technologies, the importance of these technologies for the sustainable development of the district will be reflected, without neglecting the natural beauty of the swamps located in front of the park. It will have spaces to enjoy nature, as well as the application of renewable energies. With the construction of this project, more significant economic development will be obtained. Visits to both the park and the swamps will be generated and increased, contributing and motivating the care and conservation of our environment and environmental awareness.

1. Introduction

The worldwide environmental crisis has already worried environmentalists and scientists with its most significant effect in all countries. They are affected by various factors, such as the increase in the number of the population, exploitation of natural resources, deforestation, excessive consumption of fossil fuels, and waste produced by humans [1], extensive use of fossil fuels in almost all human activities led to some undesirable phenomena, like atmospheric and environmental pollution, that had not been experienced before in known human history, in consequence, global warming, The greenhouse effect, the climate change, the depletion of the ozone layer and terminologies of acid rain began to appear in the literature frequently. These phenomena are closely related to the use of fossil fuels because they emit greenhouse gases such as carbon dioxide (CO2) [2,3], the energy park is one of the beneficial methods, educational, research, and practical use of land that is recently used in many countries. Renewable energy parks are modern parks scientific/biological/industrial / theme that would be designed and established on these concepts mentioned.

Renewable energy is the energy that comes from natural resources such as sunlight (solar energy), the wind, the rain, the tides, the hydroelectric power, and renewable geothermal heat (are replenished of the natural form) [4].

The parks are public spaces, where landscape values predominate; they are scenarios of high recreational potential and contact with nature [5]. In this sense, the parks constitute environmental protection elements, enabling healthy coexistence and social relations between members of a community.

Figure 1. Chorrillos district location

Figure 2. Chorrillos district satellite location

The Chorrillos district as to show in the figure 1, has a significant number of parks, distributed mostly irregularly, in many cases with apparent lack and deterioration, in the same way, that they have no maintenance or care by the corresponding and responsible authorities, Also, you can see the location of the project proposal can also be seen in Figure 2.

Green energy is an element key to the interactions between nature and society and is essential for economic and social development and the improvement of the quality of life worldwide [5]. Energy production is closely related to human life, affecting the industry, agriculture, and our daily life. Thus, Sustainable clean energy production can mitigate many sustainability problems facing humanity today, such as greenhouse gas emissions (GG), climate change, the depletion of fossil fuels, and energy security [2,6].

Ecological planning and design are more than a tool or technique. It is a way of mediating between human action and natural processes based on the knowledge of the reciprocal relationship between people and the earth.

2. Method

To carry out the project, we apply the following methodology consisting of two stages:

2.1 Survey of field information

At this stage, the current panorama line is characterized and thus propose ideas for implementation and improvement that allow the development of the Chorrillos district.

- Profile Measurement
- Photographic record

2.2 Ecotourism Design Development

With the previous study of the profile and the photographic record, the ecotourism design is elaborated, this consists of a theme park, whose main attraction will be the use of renewable energies, such as solar energy through panels with custom design, as well as lights posts taking advantage of this natural energy; in the same way, it will have a pool that at night will turn on lights with the energy stored during the day. All of these are distributed equally throughout the perimeter of the park. 'Smart flowers' that lean along the path of the sun, panel structures, and a photovoltaic tree will convert the light rays into electrical energy in the renewable theme park [7].

3. Results

Table 1. Environmental factors of the Chorrillos district. Environmental system

Climatic and ecological features	Average	
Temperature	Annual	28° C
	Summer	28° - 20° C
	Winter	19° - 12° C
	Spring and fall	23° – 17° C
Pluvial	6.304 in	
RH	90 – 65 %	
Cloudiness	Annual	6/7 (high)
Hours of sun	April - December	6 hours/day
	May - November	2 – 3 hours/day
Evaporation	Annual	202.6736 in
Winds	Annual speed	8.67 MPH
Atmospheric pressure	Annual	0.995 Bar
	summer	0.995 – 0.997 Bar
	winter	0.995 – 1.007 Bar

Table 2. Formatting Ultraviolet radiation levels

Risk level	UV-B index	Protection actions
Minimum	0	None.
	1	
	2	
Low	3	Apply sun protection factor.
	4	
	5	
Moderate	6	Apply sun protection factor, wearing a hat.
	7	
	8	
High	9	Apply sun protection factor, wearing a hat and glasses with UV-A and B filters.
	10	
	11	
Very high	12	Apply sun protection factor, wearing a hat and glasses with UV-A and B filters.
	13	
	14	
Extreme	>14	Apply sun protection factor, wearing a hat and glasses with a UV-A and B filter. Sun exposure for a limited time.

Table 3. Formatting Classification of pollutants according to their nature.

According to nature	Pollutants
chemical	Particulate matter
	Sedimentary
	Suspended particles
Physical or energetic	Noise
	Radiation
Biological	Greenhouse effect
	Acid rain

Source. National Service of Meteorology and Hydrology (SENAMHI).

Also, this park will have seats so that visitors can rest and enjoy the view, which will be made with materials that do not alter the landscape itself. Likewise, the proposal will have a rustic hotel, which will be of use for people who want to research the Villa Swamps and need a place to stay, being an income and recovery of the investment in the purchase of land to build the park. The climatic conditions of the area have been considered as can be seen in the table 1, table 2 and table 3 [4,8].

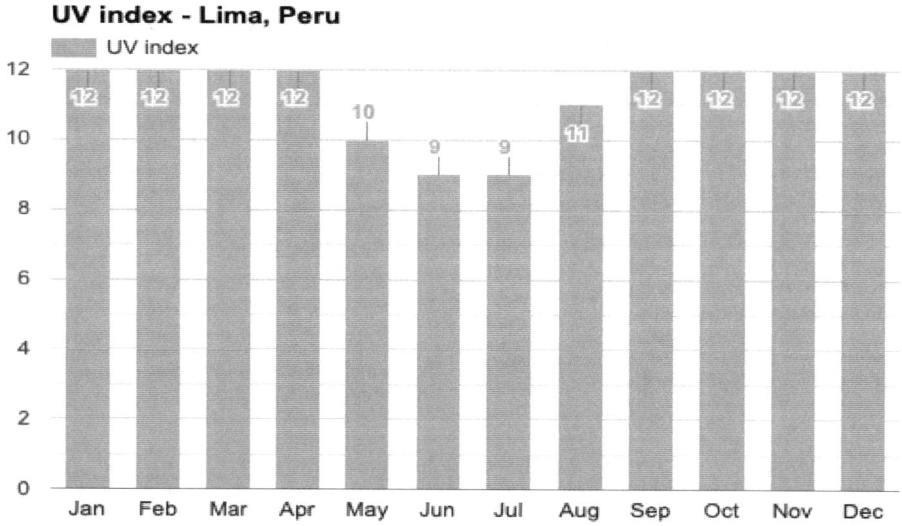

Figure 3. IUV ultraviolet radiation index, according to monthly average.Source : Weather Atlas

In figure 3 shows the performance during the year the amount of existing tradition in the study area, in which the high radiation that can be exploited can be observed.

Figure 4. Side view of the amphitheater

It can be seen in the figure 4 the simulating an entrance with stairs to the center surrounded by vegetation for the harmonization and reduction of Co2, The main lake where visitors can perform different recreational activities such as boating and enjoy nature.

3.1 Photovoltaic solar energy

Among the applications that this type of energy will have in Therme Park:

Public lighting: the posts distributed throughout the perimeter of the park, with solar panels, which will help with lighting for visitors.

Water pumping: two solar panels will power the pool and the water will be pumped with this energy.

Other applications: for palm trees with seats, which will have connections so that visitors can charge their mobile phones [9].

All this will be fed from a solar plant, where the energy will be concentrated, which will then be distributed through different points.

Figure 5. Houses supplied with energy stored in solar panels, taking advantage of radiation

In the figure 5 shows the implementation of photovoltaic panels, to take advantage of the capture of existing solar radiation in the area.

3.2 Biomass and biogas energy
Set of renewable organic matter of vegetable, animal origin, or originating from its natural or artificial transformation, it has enormous versatility, allowing it to obtain reliable and liquid or gaseous fuels through different procedures—plant or animal origin, which includes materials that come from natural or artificial transformation. Any type of biomass comes from the reaction of plant photosynthesis, which synthesizes organic substances from CO2 in the air and other pure materials, taking advantage of the sun's energy [3,7,10].

.

Figure 6. Small farm for learning and enjoyment of the community and the use of excreta as organic, playground for children of the community and visitors to the Theme Park to enjoy their stay in a playful way

As an energy resource, we have the livestock, which will be obtained from the small farm located within the park as to show figure 6. Animal dejections are the best raw material for biogas production through anaerobic fermentation. Although these residues also represent a natural fertilizer of the soil, their energy use does not affect the ecological balance, given that the effluent obtained as a product of digestion preserves the nutrients unchanged, allowing their return to the soil and eliminating, instead, potentially polluting elements

4. Discussion

The Renewable Energy Theme Park aims to be a space for collective recreation and education, as well as a future scenario of recreational and cultural activities, community, public and private, for the enjoyment of residents and visitors. Although, the document mentions the potential to produce a higher flow of tourists, it should be taken into account that this must occur under conditions of sustainability, minimizing impacts (which will always exist) and compensating those directly affected. It must also enhance the benefits (especially in terms of social inclusion and development of the swamp ecosystem).

The ability to generate electricity without emitting greenhouse gas emissions and the possibility of its decentralized installation make non-conventional renewable energies key to complying with strategies and policies related to climate change and universal access to electricity.

The high cost of investment and maintenance, complexity in construction, and not always high economic returns - which generally characterize renewable energy projects - pose an essential challenge for the development of these projects and the penetration of these technologies in the district.

5. Conclusion

At present, the number of visitors to the Villa Marsh is 30,000 people annually; with the development of the project, this figure will be increased by 45%, which will help with the economic development of the district and the residents of Chorrillos. Through the use of clean energy and the predominant vegetation, there will be a decrease in emissions of pollutant gases such as CO2, which will benefit the inhabitants and the ecosystem of "Pantanos de Villa," thus having a purer air. Moreover, the residents of the Chorrillos district will benefit from this Theme Park because they will have a place where they can enjoy nature, in a pleasant and innovative space.

6. References

[1] Ciro Rodriguez, Doris Esenarro, Pilar Ccorimanya, Fabiana Flores, Carmen Aylas & Jessica Lagos (2020). Proposal for a Sustainable Infrastructure Design (Ecolodge) in Quichas Town, Perú, "Test Engineering & Management" ISSN: 0193-4120 Page No. 9250 - 9256.

[2] Doris Esenarro, Ivana Escate, Leslie Anco, Carla Tassara, and Ciro Rodriguez (2020). Proposal for an Ecological Research Center for the Recovery and Revaluation of Biodiversity in the Town of Quichas-Lima, Peru. " International Journal of Environmental Science and Development, ISSN 2010-0264, DOI: 10.18178, Vol. 11, No. 4. 212-216

[3] Doris Esenarro, Ciro Rodriguez, Oswaldo Alfaro, Pervis Paredes / Pedro Amaya (2020). Mitigation of Greenhouse Gases (CO2) through the use of the Hybrid Engine (Hydrogen Fuel), ISSN: 0193-4120 Page No. 13448 – 13452, vol.83

[4] R.R. Heeres, W.J.V. Vermeulen, F.B. de Valle, Eco-industrial park initiatives in the USA and the Netherlands: first lessons, Journal of Cleaner Production, October 2004

[5] Angel Hsu, Carlin Rosengarten, Amy Weinfurter, Yihao Xie, Renewable Energy and Energy Efficiency in Developing Countries: Contributions to Reducing Global Emissions, The 1 Giganton Coalition, Third Report 2017

[6] UNESCO, International Workshop on Renewable Energies and Biosphere Reserves, Biospharenreservat Bliesgau, September 2017

[7] Doris Esenarro, Francisca Cabello, Pedro Amaya, and Carlos Vargas (2020). Camping Area and Dock with Viewpoint to Promote Sustainable Ecotourist in the Ticcllacocha Lagoon, Tanta-Peru, International Journal of Environmental Science and Development" ISSN 2010-0264, pag. 268-272 doi: 10.18178, Vol. 11, No. 51261

[8] E. G. Hertwich, J. Aloisi de Larderel, A. Arvesen, P. Bayer, J. Bergesen, E. Bouman, T. Gibbon, G. Heath, C. Peña, P. Purohit, A. Ramirez, S. SuhGreen, Energy Choices: The benefits, risks, and trade-offs of low-carbon technologies for electricity production. Report of the International Resource Panel, UNEP (2016)

[9] Doris Esenarro, Ciro Rodriguez, Cristian Aquije, Nicasio Obregon, Luis Anicama & Cesar Arguedas (2020). Cable Car with Water Collection for Afforestation of the Solar Hill in Chorrillos, Perú, "Test Engineering & Management" ISSN: 0193-4120 Page No. 9236 – 9242.

[10] Morand, A., R. Hennessey, J. Pittman, and A. Douglas. (2015). Linking Mitigation and Adaptation Goals in the Energy Sector: A Case Study Synthesis Report. Report submitted to the Climate Change Impacts and Adaptation Division, Natural Resources Canada, 122p.

Sensitivity Analysis of Designs of Row House Planning Influencing on Local Microclimate and Building's Cooling Energy Consumption in A Tropical City

D Jareemit[1,2,3] and M Srivanit[1,2]

[1] Faculty of Architecture and Planning, Thammasat University, Pathumthani, Thailand 12121

[2] Research Cluster in Livable Environment and Architectural Design Laboratory (LEAD-Lab)

[3] Thammasat Research Unit in Architecture for Sustainable Living and Environment.

Corresponding email: jdaranee@gmail.com

Abstract. This study aims to investigate the impact of design characteristics of row house projects on outdoor thermal conditions and the building's cooling energy consumption. The studied parameters comprise the design combinations of four street canyon orientations, building-two block shapes, four street canyon's aspect ratios (H/W), and two window to wall ratios (WWR). The study firstly performs the simulations of air temperature and mean radiant temperature across the street canyons via using ENVI-met modeling. The simulated air temperature is used as input in energy modeling to calculate cooling energy consumption in a residential unit. The Standardized Regression Coefficient (SRC) obtained from the multiple regression analysis is used to determine the significant design parameters influencing on outdoor air temperature, mean radiant temperatures, and building's cooling energy consumption. It is found that the increase of H/W has a positive effect on both outdoor conditions and building energy consumption. At the same time, the impact of street canyon's orientation and building-block shape on those issues shows an invert direction. Future studies should investigate how to optimize the design for achieving better outdoor thermal conditions and building energy efficiency.

1. Introduction

Previous studies show that the design of urban morphology can lead to urban sustainability [1–2]. It can provide benefits for both a cooling living environment [3] and building energy efficiency [4]. Several design parameters such as street canyon's aspect ratios (referred to a ratio of building height to street width), street canyon orientations, a relation of open space and building density, surface materials have been investigated in the climate-sensitive planning and building energy analysis [5–7]. However, the effects of urban planning and architectural designs on both studied issues are not well integrated. Some design parameters could considerably improve outdoor thermal conditions, but it has the worst building energy consumption. In addition, implementing all those parameters either in the simulation or design requires a significant time effort and construction cost. This study aims to understand and investigate whether the designs of the row house projects regarding planning, building geometry, and WWR can make a contribution to improving outdoor thermal conditions as well as promote energy-efficient buildings. The research results can help designers having a design strategy to create a more sustainable city.

2. Research methodology

2.1. Location selection and defined case studies
This study selects Pathumthani, located in northern Bangkok, the capital of Thailand, as a studied location. This area has the highest number of housing units owing to not only there are several significant facilities supporting people living, but it is also relatively easy to travel to Bangkok. The design characteristics of row house planning in Pathumthani, observed by previous work [6], are defined into four street canyon orientations (N–S, E–W, NE–SW, and NW–SE), four aspect ratios (0.5, 0.7, 0.9, and 1.1), and two building block shape (square and rectangular). The row house typically has two to five stories with a ratio of window to wall ratio (WWR) of 20 and 80 percent. In this study, the combination of those parameters can generate 16 design scenarios for microclimate simulation and 32 scenarios for building energy analysis.

2.2. Microclimate and energy analysis
The analysis in this study separates into two parts: 1) climate-sensitive analysis and 2) building energy consumption. The study performs the simulations of hourly outdoor air temperature (T_{air}) and mean radiant temperature (T_{MRT}) via using Envi-MET model [8]. Then air temperatures near the building surface of each design scenario are averaged and used to initiate the weather data for the energy analysis using the eQUEST model [9]. A peak cooling energy consumption in the unitary floor area (Wh/m^2/day) of a residential unit on April 26th, 2016, is calculated. The setting of the simulation model of microclimate analysis and building energy consumption are validated and compared with the field measurement and a case study from previous works [6,10].

2.3. Sensitivity analysis
For sensitivity analysis, the multiple linear regression analysis is used to examine the influential design parameters, which remarkably impact the peak cooling energy consumption. Initially, the discrete parameters, especially street canyon orientation and building block shape, have to transform into dummy variables with 0/1 coding. The study uses a standardized regression coefficient (SRC) to rank the significance of independent variables influencing independent variables. The SRC value ranges 0–1, which the value close to 1 represents the most substantial influential variables on the outputs. The SRC can be a negative value, which means the relationship between the variables has an invert direction. Finally, the influential design parameters considerably impacting on building cooling energy consumption, and outdoor thermal conditions are compared and discussed. Figure 1 summarizes the research framework of this study.

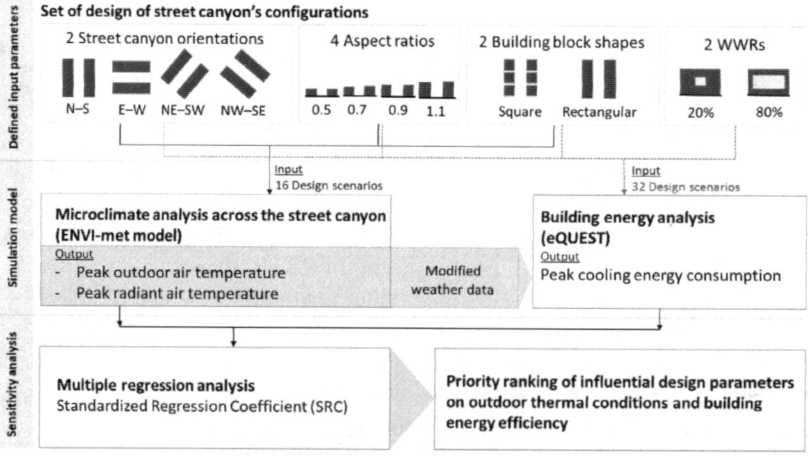

Figure 1. Research framework

3. Results and discussions

3.1. Simulations of outdoor thermal conditions

From the simulation results, the average *Tair* in the street canyon with different design configurations has a similar performance (shown in figure 2(a)). Increasing the street canyon's H/W improves the outdoor air condition slightly by up to 1%. With regards to the mean radiant temperature (in figure 2(b)), the average T_{MRT} in the street canyons with a square block shape and that of the rectangular block shows a similar performance. For a square-shaped building, the maximum T_{MRT} occurs in the N–S and E–W oriented canyons while the maximum T_{MRT} of the rectangular-shape building happens in the E–W orientation. Design with a rectangular-shaped building can lessen T_{MRT} in the N-S oriented canyons, in which the temperatures are lower than that of the canyons with a square-shaped one by 11–19 percent. Overall, an increase of street canyon aspect ratios from 0.5 to 1.1 significantly reduces the T_{MRT} by approximately 19 percent in most orientations except the E–W oriented canyons.

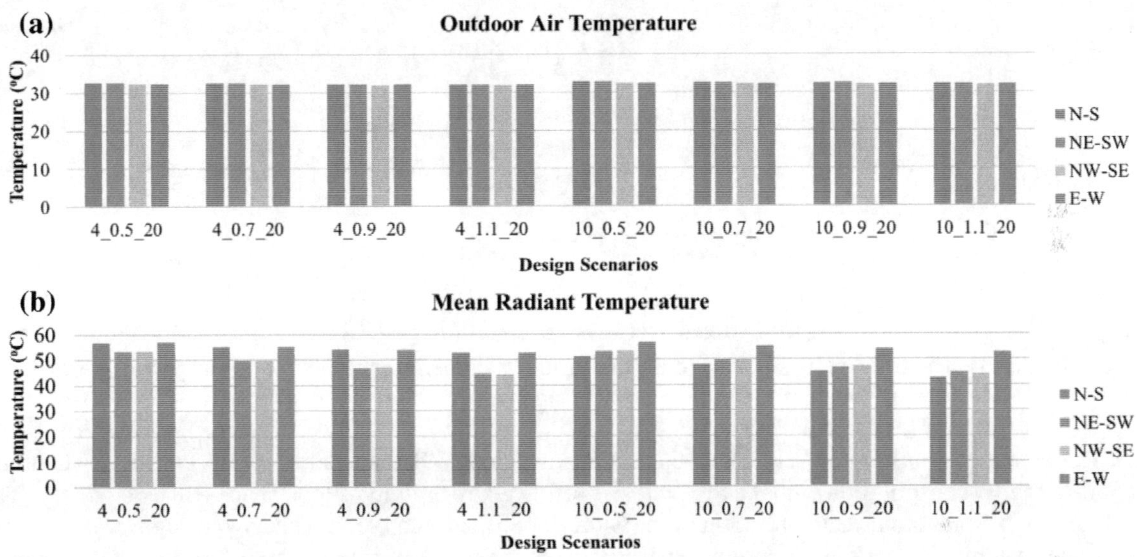

Note: For the code for 4_0.5_20, the first number represents building block shape; 4 is a square shape, and 10 is a rectangular shape. The second letter represents H/W, and the third is the WWR.

Figure 2. The simulations of average outdoor air temperature (a) and mean radiant temperature (b) across the street canyons for various design configurations

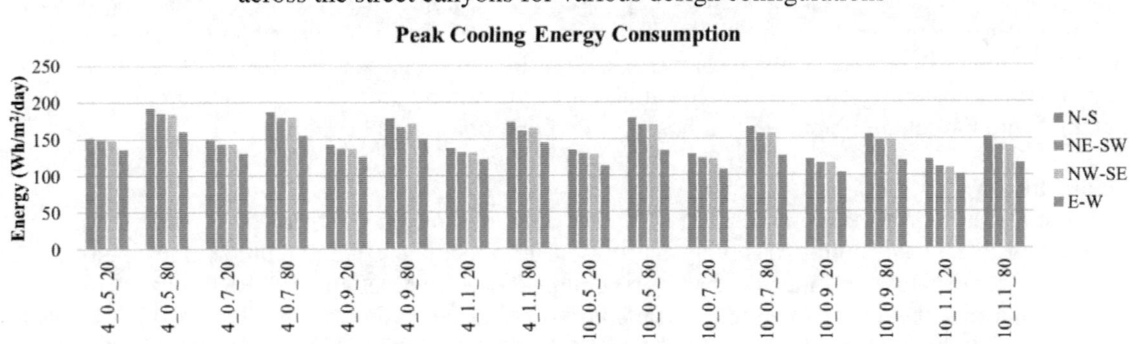

Figure 3. The simulations of the peak cooling energy consumption in the row house unit for various street canyon configurations

3.2. Calculated cooling energy consumption in a residential unit

Figure 3 presents the simulation results of cooling energy consumption in a residential unit. The row houses in the N–S oriented canyons have the highest cooling energy, while those in the E–W oriented canyons consume the lowest cooling energy. The cooling energy consumption increases as the WWR increases. The decrease of WWR significantly reduces the cooling energy consumption more than that by the increase of H/W.

3.3. Sensitivity analysis

Figure 4 presents the rank of the SRC value of each design parameter. There is a high correlation among those design parameters with the output parameters. All models have the p-value below 0.05, and the adjusted R-square values range from 0.534–0.731. The priority ranking of the influential design parameters on the T_{air} shows similar to those of T_{MRT}. The change in H/W has the highest correlation to the T_{air} and T_{MRT} (SRC = 0.619–0.759) followed by street canyon orientation (SRC = 0.282–0.330), and building-block shape (SRC = 0.196–0.230), respectively. In the climate-sensitive analysis, WWR is not accounted for as an independent design parameter according to this research scope.

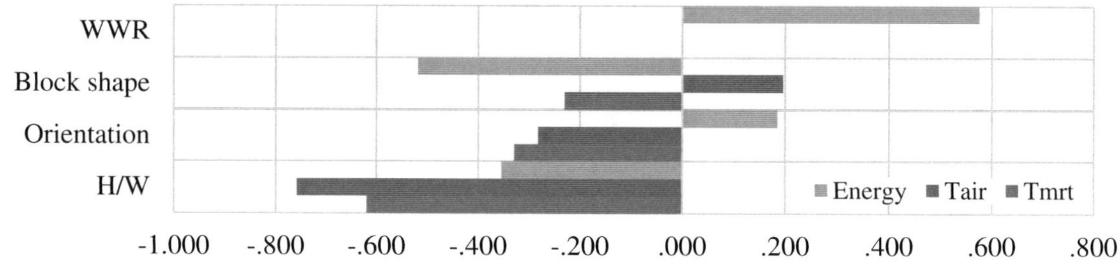

Figure 4. The SRC values determining the influential design parameters on the outdoor air temperature, mean radiant temperature, and peak cooling energy consumption.

Regarding the building energy analysis, the change in WWR considerably impacts the residential cooling energy consumption with the SRC=0.567, followed by building-block shape (SRC=0.520), H/W (SRC=0.355), and street canyon orientation (SRC=0.185), respectively. The WWR shows a positive correlation with the building's cooling energy consumption. It means that the building with higher WWR is more cooling energy consumption than that with lower WWR.

According to the sensitivity analysis, the increase of H/W could improve the outdoor thermal condition and reduce residential cooling energy consumption. However, the design of street canyon orientation and building-block shape show an invert effect on improving outdoor thermal conditions and building energy consumption. The N–S oriented canyons provide a better outdoor thermal condition than that of the E–W oriented canyons; however, the residential unit located in this orientation consumes the highest cooling energy. The energy consumption in the canyons with the rectangular block e is lower than that with the square block, while it has the worst outdoor thermal conditions.

4. Conclusion

In this study, a multiple regression method is used for assessing the set of design parameters, which make a significant contribution to cooling energy consumption in a residential unit and its impact on the outdoor air temperature and mean radiant temperature across a street canyon. Regarding the climate-sensitive analysis, the design of street canyon configurations has a significant effect on T_{MRT} across the street canyon, whereas those designs affect T_{air} slightly. The increase of H/W could provide the benefits for both outdoor thermal environment and residential energy efficiency while the design of street canyon's orientation and building-block shape give opposite results. The most significant design parameter influencing the building cooling energy consumption is WWR, whereas the H/W has the most impact on the thermal condition. For future studies, those design effects on building energy use and cooler outdoor environments for more complex urban forms should be investigated. To find the design solutions for a sustainable city, future studies should investigate and optimize the best solution enhancing both outdoor thermal condition and energy efficiency.

5. References

[1] Mutani G and Fiermonte F 2016 Microclimate Models for a Sustainable and Liveable Urban Planning Urban Planning. In: Ingaramo R, Voghera A, Topics and Methods for Urban and Landscape Design: From the river to the project. *Springer International Publishing.* **183–209** Available from: https://doi.org/10.1007/978-3-319-51535-9_11

[2] Bahgat R, Reffat RM and Elkady SL 2019 Energy Efficiency Design Guide for Optimal Urban Features of Open Spaces in Residential Complexes. *The Academic Research Community publication.* 3(1):**136-152**

[3] Deng J-Y and Wong NH 2020 Impact of urban canyon geometries on outdoor thermal comfort in central business districts. Sustainable Cities and Society. 53:**101966**. Available from: http://www.sciencedirect.com/science/article/pii/S2210670719323480

[4] Resch E, Bohne RA, Kvamsdal T and Lohne J 2016 Impact of Urban Density and Building Height on Energy Use in Cities. Energy Procedia 96:**800–814** Available from: https://doi.org/10.1016/j.egypro.2016.09.142

[5] Ahn Y and Sohn D-W 2019 The effect of neighbourhood-level urban form on residential building energy use: A GIS-based model using building energy benchmarking data in Seattle. *Energy and Buildings*, *196*, **124–133** https://doi.org/10.1016/j.enbuild.2019.05.018

[6] Srivanit M and Jareemit D 2016 Urban Planning and Environmental Design Strategies for Encouraging Transit-Oriented Development (TOD). National Research Council of Thailand

[7] Huang K-T and Li Y-J 2017 Impact of street canyon typology on building's peak cooling energy demand: A parametric analysis using orthogonal experiment. Energy and Buildings 154:**448–464**.

[8] Bruse M 1999 The influences of local environmental design on microclimate development of a prognostic numerical model ENVI-met for the simulation of wind, temperature and humidity distribution in urban structures Ph.D. thesis University of Bochum Germany

[9] Hirsch J. eQuest Introductory Tutorial, version 3.64 [Internet]. James J. Hirsch & Associates; 2010. Available from: http://doe2.com/download/equest/eQ-v3-64_Introductory-Tutorial.pdf

[10] Sudprasert S 2019 Evaluation of Energy Savings by Retrofitting of the Building Envelope of Air-conditioned Row House. *Journal of Architectural/Planning Research and Studies (JARS) 16*(1) **83–92** https://www.tci-thaijo.org/index.php/jars/article/view/152099

Acknowledgment

The authors would like to acknowledge Pichamon Leethongin and Parinthorn Canyookt for the field data collection and. We gratefully thank you for special supports from the research team of LEAD-Lab and the Faculty of Architecture and Planning, Thammasat University and from Thammasat Research Unit in Architecture for Sustainable Living and Environment.

IOP Conf. Series: Materials Science and Engineering **910** (2020) 012023 doi:10.1088/1757-899X/910/1/012023

Aggregation of Thai arborist judgments on urban tree hazard inventories used to determine tree health at single-tree level

S Kaewkhow[1] and M Srivanit[1,2]

[1] Faculty of Architecture and Planning, Thammasat University, Pathumthani, Thailand.
[2] Research Cluster in Livable Environment and Architectural Design Laboratory (LEAD-Lab)

E-mail: suppawad@gmail.com

Abstract. Tree risk assessment has evolved greatly in recent years, from ground-based visual hazard inspection methods. In this paper, we focus on Thai arborist judgment aggregations on the common optimization of urban tree hazard inventories by using the visual tree assessment (VTA) method to develop qualitative risk assessments for tree risk mitigation. The aim of this study was to develop a tree hazard inventory identifies at single-tree level based on 35 parameters with interviewing arborist experts. And then to assess parameters that are relevant to the VTA method, content validity analysis used as the basis for screening the optimal tree inventories. The results showed that VTA inventories as having 11 parameters and the definition was clarified on tree health assessment at single-tree level. Understanding the tree inventories and tree hazard level can promote management decisions that will improve public health and environmental quality in urban areas of Thailand.

1. Introduction

The VTA (Visual Tree Assessment) method has spread very rapidly world-wide and it is the basis of many cities, not least because it mainly involves visual monitoring at single-tree level, which is cost-effective and gentle on the tree. The VTA method has already been the basis of several judgements in whole or in part in such manner as it may see fit within their own city [1,2], particularly in the implementation as many large Thailand cities such as Bangkok Metropolitan Administration and Chiang Mai Municipality. The focus on information at single-tree level reflects the motives and goals of municipal authorities regarding urban forestry programs, where hazard management, traffic safety, arboricultural management, species choice, replanting decisions, and monitoring changes in the urban forest have been found to dominate [3]. However, there are several ways in which arborists might be engaged. Cities or municipal authorities in a matter may engage their own expert, and have increasingly started to perform tree inventories. In many instances, especially in urban landscape planning matters, the local authority will have engaged a number of professionals to produce the urban tree hazard reports required by council for a local environment application. These arborist specialists and their reports may end up in court if local council does not approve the tree hazard assessment and management. Thus, tree inventories at single-tree level have become valuable data sources for city authorities studying the environmental, social, and economic services provided by urban trees in Thailand. Moreover, the VTA method has limitations regarding the data parameters that can be collected at single-tree level and their measurement accuracy on tree health assessment, a fact that needs to be taken into account. In this study, the aggregation of arborist judgments by the in-depth interview technique was employed in this research. Being professional arborists involves using the VTA skills to deliver an appropriate level of tree assessment knowledge and to ensure that tree inventory parameters is accurately for non-arborist expert must fulfill. Results from this study can be used to help support which variables communities should be concerned with as they develop their local urban tree risk management programme.

2. Research Methodology

In total, 35 parameters used to determine tree health at single-tree level were identified across several studies [1,4,5,6,7,8], and grouped into 11 types of data-based parameters (e.g. age of planting, tree location, crown size and density, damage to the tree, diameter at breast height, hazard status, interaction with infrastructure or buildings, planting site, maintenance needs and history, management information, species, tree appearance and use, and vitality and health condition) (see figure 1). Then, the in-depth interview technique was undertaken with the three Thai experts of arboriculture (a professional arborist and two lecturers from landscape architectural and forestry departments) in order to elicit which tree inventory parameters reflected the tree health at single-tree level (figure 2 and figure 3). In this study, evaluate content validity will be test using the Item Objective Congruence (IOC) Index (index vary from 0.00 to 1.00) developed by Rovinelli and Hambleton (1977) [9], as the basis for screening the tree inventory parameters quality. The qualified parameters should have the IOC equal to or greater than 0.67, are considered as high reliability of urban tree hazard inventory parameters. If the IOC values which are below 0.67, will be considered as low or poor reliability.

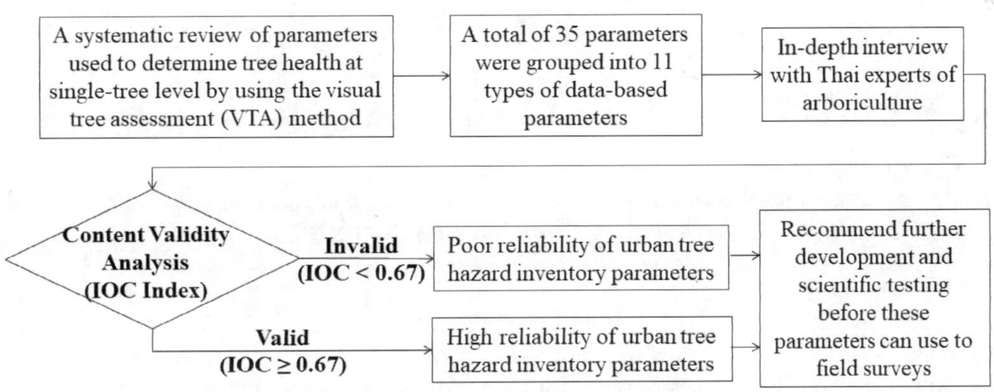

Figure 1. A proposed framework for the study of tree hazard inventory identifies at single-tree level by using the visual tree assessment (VTA) method.

Figure 2. (a) In-depth interview with Thai expert of arboriculture, and (b) a field visit to the Yang tree (*Dipterocarpus alatus* Roxb.) healthcare programs a long Chiang Mai – Lamphun route.

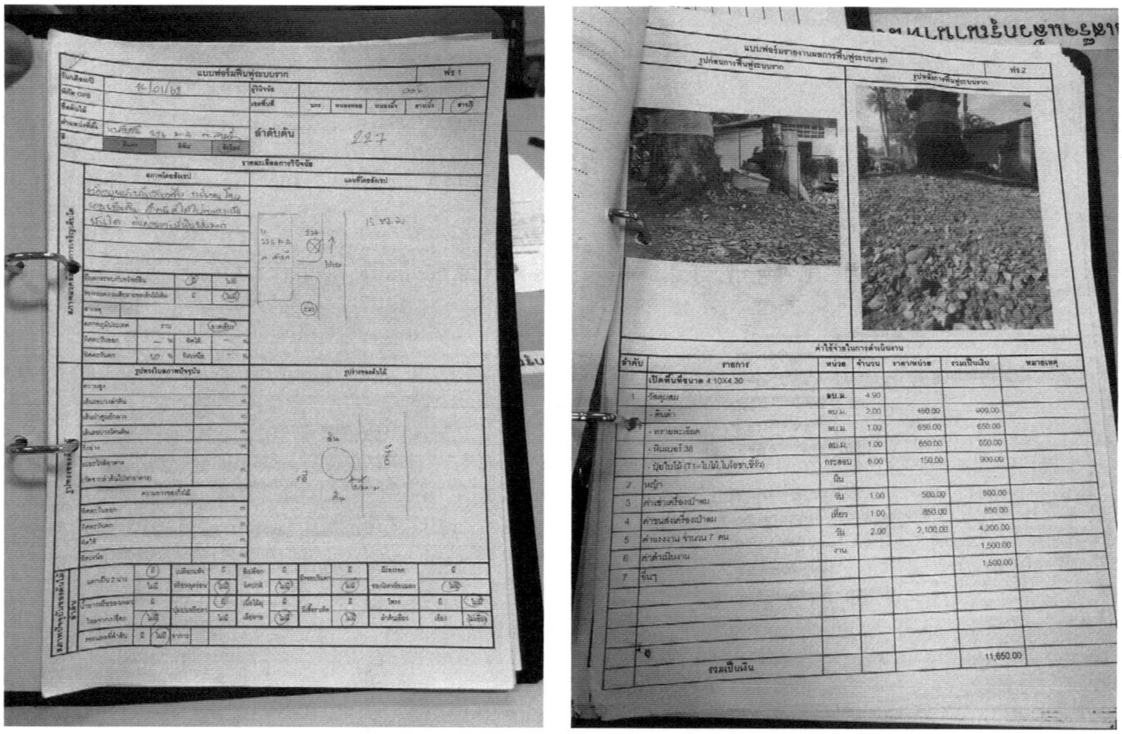

Figure 3. A sample of basic tree risk assessment form used to determine tree health at single-tree level created by Bunchong Somboonchai, a professional arborist at Faculty of Architecture and Environmental Design, Maejo University.

3. Results and Discussion

The arborist experts tried to find the more specific parameters that could be described and covered by the tree inventories. The examination of the content validity by three experts have the evaluation of content validity score by IOC values ranged from 0.67 to 1.00. An eleven group of tree parameters are provided in Table 1 as a testing for interpretation purposes.

The IOC result shows that all inventory parameters (n=35) were a valid initial test with the overall reliability of 0.76 by mean IOC. The index value of 1 for tree appearance and use indicates (a list of valid parameters include cavities and windbreak width) that all experts agreed that the item is clearly measuring objective. The damage from insects and pests also has a high index of IOC value (IOC=0.86), indicating that the experts agree these parameters are clearly using to determine tree health at single-tree level (including damage to the tree, dieback, discolored leaves, mechanical damage, tree mortality, presence/absence of chlorosis, and presence of insects/disease). Only one of three experts believed that it is unclear whether the tree parameter (IOC=0.67) is the measure of age or year of planting, crown size and density, DBH, planting site, maintenance needs and history.

A five group of tree parameters were suggested by experts for improvement measures to determine tree health at single-tree level, as those parameters that had a similar content to other items, even though the IOC score of those items was more than 0.67. Moreover, to be useful in risk assessment and management, inventories must be regularly updated to include plantings, removals, and changes in tree condition. Obviously, this includes the collection of data that goes beyond what is needed to conduct a basic tree health at single-tree level. In contrast, the presence of an actively updated inventory was a key differentiation, which helped predict which programmes regularly conducted tree risk management [10].

Table 1. Index of items objectives congruence of types of tree data-based parameters

Group of tree parameters	Inventories (n=35)	Mean of IOC	IOC*	Determine which valid/invalid parameters: Valid (IOC ≥ 0.67)	Invalid (IOC < 0.67)
Age or year of planting	1	0.67	0.67	Number of years after transplantation	None
Crown size, density	5	0.67	0.67	Crown size	None
			0.67	Canopy transparency	
			0.67	Crown damage	
			0.67	Crown density	
			0.67	Crown dieback	
Damage, insects, and pests	7	0.86	1.00	Damage to the tree	None
			0.67	Dieback	
			1.00	Discolored leaves	
			1.00	Mechanical damage	
			0.67	Tree mortality	
			0.67	Presence/absence of chlorosis	
			1.00	Presence of insects/disease	
DBH (diameter at breast height)	1	0.67	0.67	Diameter at breast height	None
Hazard status	2	0.84	0.67	Existence of girdled roots	None
			1.00	Tree part most likely to fail	
Interaction with infrastructure or buildings	5	0.74	1.00	Proximity to building	None
			0.67	Building direction	
			0.67	Conflict with infrastructure	
			0.67	Whether the sidewalk is raised	
			0.67	Tree position in relation to traffic	
Location	6	0.78	0.67	Available planting spaces	None
			0.67	Location	
			1.00	Geographical location	
			1.00	Local name	
			0.67	Location of the trees	
			0.67	Street name	
Planting site	3	0.67	0.67	Amount of impermeable cover	None
			0.67	Land use type	
			0.67	Site condition	
Maintenance needs and history	1	0.67	0.67	Pruning	None
Size (except crown and DBH)	2	0.84	0.67	Number of trunks	None
			1.00	Height	
Tree appearance and use	2	1.00	1.00	Cavities	None
			1.00	Windbreak width	

Note: **summary of three professional arborist experts' judgments regarding tree parameters of VTA.*

4. Conclusions

Because there are no statistical tests for assessing significance of the measure, a substantive procedure is recommended for setting criterion levels. For example, an index value of 0.67 would occur if two thirds of the decisions had perfect agreement that a parameter is measuring what it has been defined to measure and one thirds of the professional arborist experts were unsure if the item is measuring each of the objectives. The study recommended inspection of the objective averages for more detailed diagnostic purposes. An IOC value of 0.67 when two of three experts have perfect agreement on the objective being measured may be acceptable as other parameters. However, the IOC indicated that having just some form of inventory was not sufficient as the more ambiguous tree inventory parameter was used to determine tree health at single-tree level for urban tree conditions in Thailand.

Findings from this study identify the current state of urban tree risk management in the local or municipal authorities in Thailand, as well as the tree inventories which either cause or are associated with tree hazard assessment and management, and an investment in the collection and analysis of relevant urban tree data, to make informed decisions in their urban areas. Ideally, a municipal programme designed to address tree risk management should be developed in a proactive manner, before an incident involving a tree-related failure occurs. Once implemented, the programme should help to reduce the risk associated with personal injury or property damage.

5. References

[1] Schipperijn J, Pillmann W, Tyrvainen L, Mäkinen K and O'Sullivan R 2005 Information for urban forest planning and management *Urban forests and trees* Springer The Netherlands 399–417
[2] Nielsen AB, Östberg J and Delshammar T 2014 Review of Urban Tree Inventory Methods Used to Collect Data at Single-Tree Level *Arboriculture & Urban Forestry* 40(2) 96–111
[3] Keller JKK and Konijnendijk CC 2012 Short communication: A comparative analysis of municipal urban tree inventories of selected major cities in North America and Europe *Arboriculture & Urban Forestry* 38 (1) 24–30
[4] Matheny N and Clark J 1994 A photographic guide to the evaluation of hazard trees in urban areas (2nd ed.) Champaign IL *International Society of Arboriculture.*
[5] Fink S 2009 Hazard tree identification by visual tree assessment (VTA): Scientifically solid and practically approved *Arboricultural Journal* 32 139–155
[6] Norris M 2007 Tree risk assessments – what works – what does not – can we tell? A review of a range of existing tree risk assessment methods *In ISAAC Conference Proceedings* Perth p 29
[7] Paine LA 1971 Accident hazard evaluation and control decisions on forested recreation sites (Res. Paper PSW-68) *Berkeley CA: Pacific Southwest Forest & Range Experiment Station* Forest Service U.S. Department of Agriculture
[8] Östberg J, Wiström B and Randrup TB 2018 The state and use of municipal tree inventories in Swedish municipalities – results from a national survey *Urban Ecosystems* 21 467–477
[9] Rovinelli RJ and Hambleton RK 1977 On the use of content specialists in the assessment of criterion-referenced test item validity *Dutch Journal of Educational Research* 2 49-60
[10] Koeser AK, Hauer RJ, Miesbauer JW and Peterson W 2016 Municipal tree risk assessment in the United States: Findings from a comprehensive survey of urban forest management *Arboricultural Journal* 38(4) 218-229

Acknowledgments

This study was supported by Thammasat University Research Fund, Contact No. TUGT 2/2562. The authors would like to express a heartfelt thanks to the involved professional arborists include Taradon Tunduan, Bunchong Somboonchai and Dr.Ponthep Meunpong, for their participation and contribution in the investigation. We especially want to thanks Khwanchanok Deepatthana and the team of the LEAD-Lab for assisting this research project.

IOP Conf. Series: Materials Science and Engineering **910** (2020) 012024 doi:10.1088/1757-899X/910/1/012024

Public spaces in Bangkok and the factors affecting the good public space quality in urban areas

C Kongphunphin[1] and M Srivanit[1,2]

[1] Faculty of Architecture and Planning, Thammasat University, Pathumthani, Thailand.
[2] Research Cluster in Livable Environment and Architectural Design Laboratory (LEAD-Lab)

Corresponding author's e-mail address: chompooploy2012@gmail.com

Abstract. Public spaces are an important component of the urban area used to support public activities. The aim of this study is to evaluate the quality of public spaces where subdivided into districts of Bangkok city area. A methodology is carry out preliminary site surveying in a 31 public parks based on four main good public space indicators, which including accessibility, imageability, usability and sociability. In this study, the Z-score or standard score is defined as a statistical measurement of the associate of site characteristics score, has been apply to determine quality of public spaces. The results reveal that the all of public spaces are outstanding scores in terms of imageability and usability, and their Z-score values are consistently coherent, especially the high-quality public spaces located in the inner zone of Bangkok. While accessibility and sociability factors are poor quality, their values show on some public spaces that mostly located in the middle and in the outer zones of Bangkok. These results can be helpful to given a priority when managing city facilities in the Bangkok's urban space and maintaining or improving its public space value.

1. Introduction

Public spaces are one of the most important of social components of the city. They have been used to support human activities in daily life from the past to present [1,2,3]. Within the public spaces, there are the various relationship of interaction between people, people and space, and its surrounding [4,5]. Public spaces associate with their neighbourhoods and located districts, reflecting social, economic and environment [5,6]. They were shaped and shaping together, coursing the unique character of them both public spaces and their located zone [6,7]. Quality of a city is often associated with its public spaces. Good public spaces give many advantages to all sectors of a city [8]. The standard of public spaces has no stereotype, they were assessed their quality by many indicators with vary scholars. Good public space began to be considered as an important issue by Jan Gehl since 1974 [9] and it has been continuously consideration on urban planning to the present. There are many major terms for improving quality of public spaces. Moreover, there are many keywords for good design of public open spaces including protection, comfort and enjoyment, cleanliness, tidiness, accessibility, attractiveness, comfort, inclusiveness, vitality and viability, function, distinctiveness, safety and security, robustness, greenness, unpollutedness and capacity for fulfilment [10], liveability, character, connection, mobility, personal freedom and diversity [11], access and linkages, comfort and image, uses and activities and sociability [12], imageability, visual enclosure, human scale, transparency and complexity [13]. This study is a preliminary survey of urban public space in Bangkok, the city was divided to three zones according to the location of the area [14]. As shown in figure 1, the three circles represent the different urban development density zones of Bangkok.

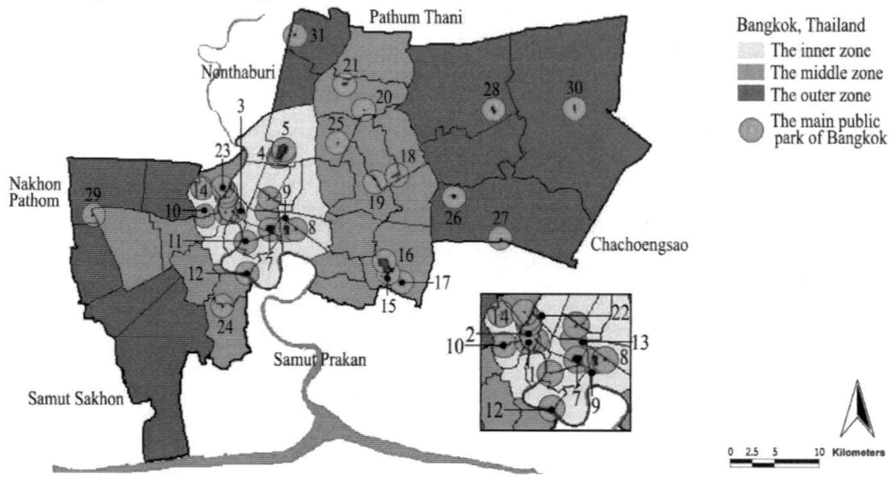

Figure 1. The location of public parks in three areas represent the different urban development density zones of Bangkok.

2. Research Methodology

In order to get the results, this study used a checklist of elements of public space covered four dimensions and thirteen factors as seen in table 1. In terms of accessibility, there are two factors refer to convenience of entering including number of bus station point and entrance. As regards imageability, it is the physical relationship of the surrounding environment with each other, which creates images of the distribution of urban parks and understanding for the users within the service areas – a walking distance of 500 meters from any parks. There are four factors represent to convenience, importance, background or history and appreciation consisting of the number of facilities, age of park, be donated from aristocratic and wealthiest people (refer to an important history of parks), classify the park contents of the city images into five types of elements which are paths, edges, districts, nodes and landmarks. Regarding usability there are five factors that interpret to role and usage of public spaces inclusive of the number of user on weekday and weekend, size, the number of role and the number of important building which located on the public spaces. With regard to sociability, there are two factors that citing to social significance of public spaces which includes the number of nation of user and element of social (in case of Thai context), which including residence in background, educational institution and religious institution. On the process of analysis, all factors' scores were converted to standard score or Z-score, a unit of measurement that is converted from raw scores to measurement results, used to describe the quality level of public spaces in each zones. The formula for calculating is given below:

$$Zi = \frac{Xi - \bar{X}}{S} \quad (1)$$

Where, Xi is a raw score, $\bar{X}$ is the mean of the sample and S is the standard deviation of the sample.

The Z-score can be both plus or minus, showing that the Z-score's ability is above or below average respectively. Each Z-score of the factors and dimensions were considered on comparison among thirty-one main parks.

IOP Conf. Series: Materials Science and Engineering **910** (2020) 012024 doi:10.1088/1757-899X/910/1/012024

Table 1. Summary of dimensions/factors and Z-scores affecting the good public spaces quality in Bangkok urban areas (N=31)

Zoning	Park	Dimensions/Factors												
		Accessibility			Imageability			Usability				Sociability		
		Bus station	Entrance	The number of	Age of park	Be donated by important	The number of elements of image of the	The number of users on	The number of users on	size	No. of roles	The number of	The number of nation	The number of
	1 Rommaninat Park	-0.083	-0.141	-0.178	0.167	-0.379	-0.205	-0.114	-0.356	-0.508	-0.268	1.867	-0.258	0.083
	2 Saranrom Palace Park	1.073	0.268	-0.002	2.164	**2.556**	-0.092	0.976	1.097	-0.555	-0.050	**4.512**	-0.258	-0.161
	3 Santi Chai Prakan Public Park	0.353	2.160	0.542	-0.192	-0.379	1.318	**4.297**	**4.507**	-0.672	1.071	1.662	-0.258	1.219
	4 Chatuchak Park	-0.135	-0.479	-0.509	1.140	**2.556**	-0.536	-0.355	0.610	0.474	-0.505	-0.325	-0.258	-0.565
	5 Wachirabenchathat Park	-0.468	-0.679	-0.556	-0.294	-0.379	-0.603	-0.252	0.077	2.188	-0.611	-0.325	-0.258	-0.504
	6 Queen Sirikit Park	-0.515	-0.564	-0.512	0.013	-0.379	-0.573	0.129	-0.401	0.793	-0.577	-0.325	-0.258	-0.516
	7 Lumphini Park	-0.455	-0.596	-0.500	**3.650**	**2.556**	-0.610	-0.173	-0.218	2.071	-0.594	-0.325	-0.258	-0.564
Outer	8 Benchasiri Park	0.275	-0.341	-0.021	0.218	-0.379	-0.313	1.663	1.228	-0.508	0.016	-0.325	-0.258	-0.247
	9 Benjakitti Park	-0.581	-0.652	-0.468	-0.499	-0.379	-0.564	-0.551	-0.501	0.279	-0.563	-0.325	-0.258	-0.504
	10 Parco Sirintorapusapan	1.111	**3.127**	**4.466**	-1.062	-0.379	**3.534**	1.445	1.102	-0.711	**3.935**	-0.325	-0.258	2.616
	11 Chaloem Phrakiat Forest Park	-0.497	-0.189	0.271	-0.806	-0.379	-0.041	0.078	0.060	-0.570	0.007	-0.325	-0.258	-0.122
	12 Public Park in Commemoration of H.M the King's	0.006	-0.341	-0.230	-0.192	-0.379	-0.097	0.061	0.108	-0.508	-0.173	-0.325	-0.258	-0.082
	13 Santiphap Park	1.327	0.999	-0.141	-0.089	-0.379	0.145	0.498	0.267	-0.578	0.040	-0.325	-0.258	0.141
	14 Public Park in Commemoration of the King Rama	0.449	-0.059	0.118	-0.755	-0.379	0.099	-0.496	-0.282	-0.602	0.161	-0.325	-0.258	0.268
	15 Chaloem Phrakiat Mahat Thai Park	-0.361	-0.161	-0.065	-0.550	-0.379	-0.167	0.023	-0.106	-0.578	0.178	-0.325	-0.258	-0.338
	16 King Rama IX Park	-0.595	-0.660	-0.566	0.474	-0.379	-0.611	-0.557	-0.452	**3.162**	-0.620	-0.325	-0.258	-0.568
	17 Wanatham Park	0.044	-0.606	-0.350	-0.294	-0.379	-0.345	-0.664	-0.715	-0.399	-0.200	-0.325	-0.258	-0.466
	18 Serithai Park	-0.576	-0.542	-0.558	-0.038	**2.556**	-0.601	-0.714	-0.769	1.993	-0.600	-0.325	-0.258	-0.564
	19 Nawamin Phirom Park	-0.518	-0.512	-0.497	-0.704	-0.379	-0.472	-0.551	-0.540	-0.142	-0.502	-0.325	-0.258	-0.515
Middle	20 Watcharapirom Park	-0.239	-0.400	-0.418	-0.960	-0.379	-0.268	-0.286	-0.207	-0.469	-0.324	-0.325	-0.258	-0.296
	21 Ramintra Sports Park	-0.533	-0.446	-0.391	0.474	-0.379	-0.424	-0.502	-0.473	-0.274	-0.414	-0.325	**3.746**	-0.415
	22 Rama VIII Park	-0.231	-0.161	-0.065	-0.294	-0.379	0.145	-0.337	-0.322	-0.578	-0.097	-0.325	**3.746**	-0.099
	23 Charan Pirom	**4.573**	**3.127**	2.441	-0.960	-0.379	3.534	-0.558	-0.607	-0.711	3.019	-0.325	-0.258	**4.213**
	24 Thonburirom Park	-0.456	-0.465	-0.332	1.447	-0.379	-0.438	-0.223	-0.360	-0.243	-0.429	-0.325	-0.258	-0.350
	25 Ladprao water park	-0.497	-0.189	-0.091	-0.755	-0.379	-0.041	-0.747	-0.754	-0.570	-0.124	-0.325	-0.258	-0.122
	26 Garden 60th Anniversary Queen	-0.493	-0.266	-0.323	-0.601	-0.379	-0.431	0.540	0.426	-0.259	-0.467	-0.325	-0.258	-0.342
	27 Suan Ladkrabang (Suan Pranakorn)	-0.465	-0.277	-0.323	0.474	-0.379	-0.324	-0.449	-0.425	-0.344	-0.372	-0.325	-0.258	-0.290
Inner	28 Wareephirom Park	-0.513	-0.597	-0.521	-0.909	-0.379	-0.533	-0.652	-0.588	0.209	-0.534	-0.325	-0.258	-0.538
	29 Thawee Wanarom Park	-0.525	-0.311	-0.344	-0.294	-0.379	-0.405	-0.607	-0.582	-0.313	-0.342	-0.325	-0.258	-0.400
	30 Nong Chok Public Park	-0.547	-0.078	-0.293	0.474	-0.379	-0.279	-0.330	-0.103	-0.461	-0.333	-0.325	-0.258	-0.030
	31 Rommani Thungsikan Park	0.072	0.032	0.416	-0.448	-0.379	0.198	-0.592	-0.721	-0.617	0.269	0.735	-0.258	0.061
	Mean	**149.38**	**79.845**	**245.590**	**21.742**	**0.129**	**127.178**	**24407.181**	**30535.237**	**0.151**	**147.12**	**12.781**	**1.065**	**75.376**
	Standard Deviation	**240.66**	**107.722**	**411.507**	**19.524**	**0.341**	**199.831**	**30685.468**	**36556.657**	**0.205**	**227.33**	**39.323**	**0.250**	**130.457**

Highest score Lowest score

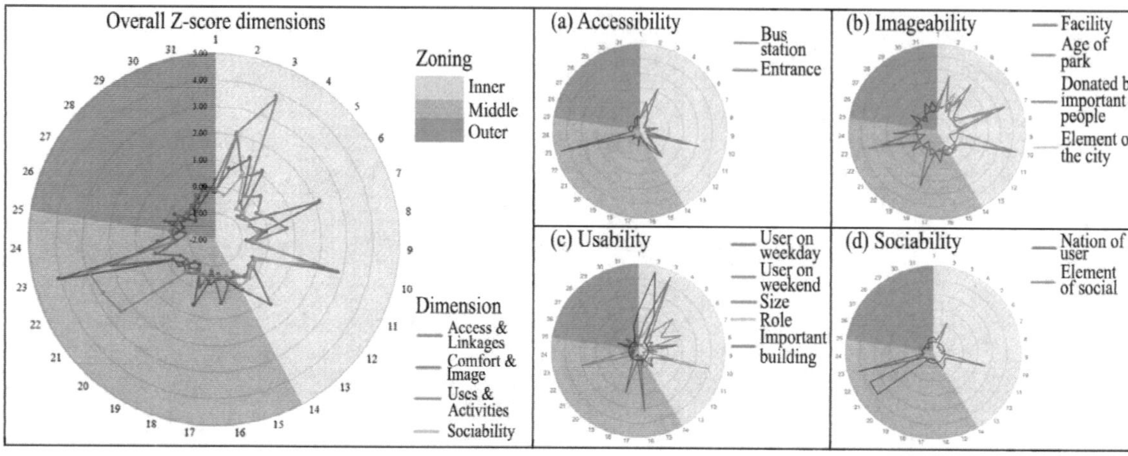

Figure 2. The overall Z-score dimensions of public spaces in different urban zones and the Z-scores of accessibility (a), imageability (b), usability (c) and sociability (d).

3. Results and Discussion

The Z-score of the factors of each dimension of thirty-one public spaces among three zoning of Bangkok are various, every Z-score imply to performance and quality of public spaces where located on each zoning which can be seen in table 1. A comparative performance of the public spaces in three zones was represented in the form of a spider diagram which can be seen in figure 2

3.1. Public space in the inner zone of Bangkok

The quality of public space in the inner zone are the greatest than public spaces in the middle and outer zone. Their dimension Z-score are apparently high, most of them are plus status. In terms of accessibility, entrance factor is the top of all. Its Z-score is 3.127. In the same time with imageability, the number of facility, age of park, be donated by important people and the number of element of image of the city factors are the highest. Their Z-score are 4.466, 3.650, 2.556 and 3.534 respectively. And usability, the number of user on weekday and weekend, role and important building factors are the highest. Their Z-score are 4.297, 4.507, 3.935 and 4.512 respectively.

3.2. Public space in the middle zone of Bangkok

Public spaces in this zone have the unique character. Their Z-score are plus and minus status in the same dimension. In terms of accessibility, both bus station points and important people factors are the highest. Their Z-score are 4.573 and 3.127 respectively. On usability, size factor is the highest. Its Z-score is 3.162. And sociability, the number of nations of user and element of social factors are the highest. Their Z-score are 3.746 and 4.213 respectively.

3.3. Public space in the outer zone of Bangkok

The quality of public spaces in the outer zone are the poorest than the other zones. Most of their Z-score are minus. Their dimension Z-score are apparently low in every factor of the dimensions. Especially imageability, usability and sociability dimensions. Their Z-score of be donated by important people, the number of important building and nation of user are 0.379, 0.325 and 0.258 respectively.

4. Conclusion

Quality and character of public spaces in each urban zoning are different, their quality followed to physical of urban zoning. In the inner zone of Bangkok, their public spaces were used to many roles. Its public spaces are quite good quality. They perfect with good access and have many mode choices of transportation, linkage between building or place, and sight attraction from elements in its public spaces. Public spaces in the inner zone are plenary consistent with the area where there is prosperity as the districts within the city centre. While the middle zone of Bangkok is semi-suburb area, the main role of its public spaces was used to support community activities. The public spaces are various characteristics, outstanding with size; there are both the biggest and smallest area of all the main public spaces in Bangkok, convenience on access and linkages and importance in history. Moreover, their public parks are significant in sociability, they located on the districts that contain many nations of user distribute surrounded with elements of social structure of Thai. The contexts between public spaces and middle zone represents the neighbourhood that is community. And the outer zone of Bangkok is a suburb area, where their facilities are less than urban centre. Their public spaces are quiet poor, and the poorest in many dimensions including imageability, usability and sociability. The results of this study are obtained from the analysis that focused on physical factors. For more clear results, economic and social factors should be considered in further studies.

5. References

[1] Kaplan, S and Kaplan, R 1982 *Cognition and Environment: Functioning in An Uncertain World.* 6th Edition. New York: Preager

[2] Ben, SI Gaus, GF1983 *Public and Private in Social Life London:* Croom Helm Blackmar, E 2006 *Appropriating the commons: The tragedy of property rights discourse* New York: Taylor & Francis

[3] Madanipour, A 2010 *WHOSE PUBLIC SPACE?* New York: Routledge Kongphunphin, C, Iamtrakul, P & Srivanit, M 2018 The Attitude in Urban Planning of Thai Urban Public Space. *International Journal of Building, Urban, Interior and Landscape Technology.* 12*(2)*, 59-74

[4] Carmona, M 2010 Contemporary public space, part two: classification *Journal of Urban Design*, 15*(2)*, 157-173

[5] Barton, H, Grant, M and Guise, R 2010 *Shaping Neighbourhoods: For Local Health and Global Sustainability* (2nd Edition) New York: Routledge

[6] Cackowski, J M & Nasar, J L 2003 The restorative effects of roadside vegetation implications for automobile driver anger and frustration *Environment and Behaviour 35(6)* 736-751

[7] Gehl J 1987 *Life between Buildings* New York: Van Nostrand-Reinhold

[8] Carr, S 1992 *Public space* Cambridge University Press

[9] Smith, T, Nelischer, M, & Perkin, N 1997 Quality of an urban community: a framework for understanding the relationship between quality and physical form. *Landscape and Urban Planning*, 32*(2)*, 229-241

[10] Project for Public Space 2000 *How to turn a place around: a handbook for creating successful public spaces* Project for Public Spaces In corporates

[11] Ewing, R, & Clemente, O 2013 *Measuring urban design: Metrics for livable places* Island Press

[12] Department of Public Works and Town & Country Planning, 1999 *Town planning during the reign of King Bhumibol Adulyadej* Bangkok: Seven Printing Group Company Limited

Acknowledgments

The authors thank the Bangkok Metropolitan Administration for share their data and supported this study. This research did not receive any specific grant from funding agencies in the public, commercial, of not-for-profit sectors.

AUTHOR INDEX

Institute of Physics
Dirac House, Temple Back
Bristol BS1 6BE UK

CURRAN ASSOCIATES INC.
proceedings.com

ISSN: 1757-8981
ISBN 978-1-7138-3165-5

9 781713 831655